Polymers and Composites Manufacturing
Advanced Composites

Also of interest

Series: Advanced Composites.
J. Paulo Davim (Ed.)
ISSN 2192-8983
Published titles in this series:
Vol. 10: Biodegradable Composites (2019) Ed. by Ed. by K. Kumar, J. P. Davim
Vol. 9: Wear of Composite Materials (2018) Ed. by J. P. Davim
Vol. 8: Hierarchical Composite Materials (2018) Ed. by K. Kumar, J. P. Davim
Vol. 7: Green Composites (2017) Ed. by J. P. Davim
Vol. 6: Wood Composites (2017) Ed. by A. Alfredo, J. P. Davim
Vol. 5: Ceramic Matrix Composites (2016) Ed. by J. P. Davim
Vol. 4: Machinability of Fibre-Reinforced Plastics (2015) Ed. by J. P. Davim
Vol. 3: Metal Matrix Composites (2014) Ed. by J. P. Davim
Vol. 2: Biomedical Composites (2013) Ed. by J. P. Davim
Vol. 1: Nanocomposites (2013) Ed. by J. P. Davim, C. A. Charitidis

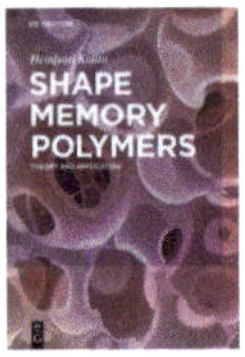

Shape Memory Polymers
Kalita, Hemjyoti, 2018
ISBN 978-3-11-056932-2, e-ISBN 978-3-11-057017-5

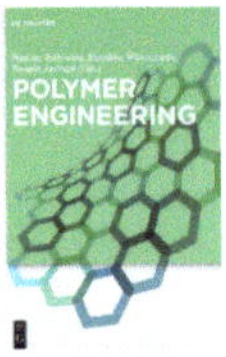

Polymer Engineering
Tylkowski, Wieszczycka, Jastrzab, 2017
ISBN 978-3-11-046828-1, e-ISBN 978-3-11-046974-5

Polymers and Composites Manufacturing

Edited by
Kaushik Kumar and J. Paulo Davim

DE GRUYTER

Editors
Dr. Kaushik Kumar
Department of Mechanical Engineering
Birla Institute of Technology
Mesra, Ranchi, Jharkhand 835215
India

Prof. Dr. J. Paulo Davim
Dept. of Mechanical Engineering
University of Aveiro
Campus Santiago
3810-193 Aveiro
Portugal

ISBN 978-3-11-065193-5
e-ISBN (PDF) 978-3-11-065504-9
e-ISBN (EPUB) 978-3-11-065212-3
ISSN 2192-8983

Library of Congress Control Number: 2019952020

Bibliographic information published by the Deutsche Nationalbibliothek
The Deutsche Nationalbibliothek lists this publication in the Deutsche Nationalbibliografie; detailed bibliographic data are available on the Internet at http://dnb.dnb.de.

Cover image: gettyimages/thinkstockphotos, Abalone Shell
Typesetting: Integra Software Services Pvt. Ltd.
Printing and binding: CPI books GmbH, Leck

www.degruyter.com

Preface

The editors are pleased to present the book *Polymers and Composites Manufacturing* under the book series *Advanced Composites*. The chosen book title reflects the upcoming trends in manufacturing of polymers and composite materials for the next decade. This book is a compilation of different aspects of the same.

One of the most outstanding features of plastics and composites is the ease with which they can be processed. In some cases, semifinished articles such as sheets or rods are produced and subsequently fabricated into a shape using conventional methods such as welding or machining. In the majority of cases, however, the finished article, which may be quite complex in shape, is produced in a single operation. There is a wide range of processing methods, which are currently being used for plastics and composites. In many cases, the choice of method is based on the shape of the component and whether it is thermoplastic, thermosetting, filler or fiber reinforcement and the cost incurred and most important of all the desired properties, as the properties of polymer and composites vary drastically with change in processing techniques.

The aim of this book is to provide a forum for researchers and practitioners to review the recent advances in the area of polymer and composite manufacturing or processing and identify possible trends for further developments as well as to attract industrial partners interested in application development with a new dimension. The advancement can be visualized with *6M* theory:

1M – Man, that is, advancement in automation, which would enhance the human–machine interaction.

2M – Method, that is, advancement in existing methods or development of a new method for ease in manufacturing, better quality, less cost and more productivity.

3M – Machine, that is, advancement in existing machines or development of a new machine for ease in manufacturing, better quality, less cost and more productivity.

4M – Mold, that is, advancement in existing molds and dies or development of a new mold and dies for ease in manufacturing, better quality, less cost and more productivity.

5M – Material, that is, advancement in existing materials or development of new materials for ease in manufacturing, better quality, less cost and more productivity. In this case, materials can be plastics, composites or materials for the mold and dies.

6M – Modeling, that is, advancement in existing software or development of new software for virtual simulation of the manufacturing process for ease in manufacturing, better quality, less cost and more productivity.

At present, looking at the wide application an insight about *Polymers and Composites Manufacturing* is the key for major discipline, and many researchers and scholars are

https://doi.org/10.1515/9783110655049-202

working in these areas. This book provides an insight for all researchers, academicians, postgraduate or senior undergraduate students working in the area. The chapters in the book have been provided by researchers and academicians working in the field and have gained considerable success in the field.

For the ease of the readers, chapters in the book have been categorized in *five sections*, namely, *Section I: Composite and Mold Design*; *Section II: Characterization and Properties*; *Section III: Simulation and Experimentation*; *Section IV: Optimization*; and *Section V: Environmental Issues.*

Section I contains *Chapters 1* and *2*, whereas *Section II* has *Chapters 3* and *4*, *Section III* has *Chapters 5–8*, *Section IV* has *Chapters 8* and *9* and *Section V* has *Chapter 10*.

Section I starts with *Chapter 1* and provides insight about design, optimization and manufacturing of monocomposite carbon/epoxy leaf spring. A steel leaf spring of light commercial vehicle is replaced by monocomposite leaf spring for weight optimization as a main objective. The same has been simulated using a simulation software and also manufactured by a very well-known technique, that is, hand layup method with hot molding process. Considerable weight reduction was achieved with same mechanical properties of the steel counterpart.

Chapter 2 shows the dependence of mold design on the fiber orientation in the case of injection molding, one of the most important processes to manufacture plastic parts, especially for complex geometries. In this chapter, parameters affecting the fiber orientation such as the mold design (especially cavity shape and gate position) during the injection molding were detailed. Then, to predict the fiber orientation during the shaping phase, various software such as REM 3D software and Moldflow insight were presented to solve the problems related to the injection of composite and finally several models of the composite mechanical performance prediction have been identified.

Section II initiates with *Chapter 3*, which discusses about characterization of an unsaturated polyester resin for liquid composite molding (LCM) processes. The process, one of the current evolutionary ones, involves different phenomena such as resin flow, heat transfer and polymerization reactions simultaneously. Mold filling and subsequent curing are the significant processing stages required to be modeled in LCM process simulation. In this process, resin polymerization reaction leads to phase transformation from viscous liquid to rigid solid with an exothermal effect; hence, gel time marks the onset of viscous resin liquid to gel stage, which constitutes a crucial parameter for the mold fill time. The required component geometry needs to be filled before the resin gels. Hence, the models depicting resin cure reaction kinetics and viscosity as a function of temperature and degree of cure constitute the submodels to the main flow, heat and mass transfer models. In this chapter, general-purpose unsaturated polyester resin was characterized for gelation and exotherm behavior, resin cure kinetics, resin cure viscosity for applications in LCM process simulations.

Chapter 4 presents a review of researches undertaken on oil palm lignocellulose fiber and polymer composite in last two and half decades. The objectives were to dilate on the effects of oil palm fiber (OPF), involving empty bunch fiber and oil palm mesocarp fiber on selected polymer materials, that is, polyethylene (PE), polypropylene, polyvinyl chloride, polystyrene, polyurethane, polyester and epoxy. The chapter mainly considered articles relating physical, mechanical, thermal and electrical characteristics of OPF-polymer blends. It was observed that an OPF-polymer-based composite material exhibited different characteristics based on five factors, namely, fiber type (bunch or mesocarp), fiber size, fiber percentage weight or volume in composite, fiber percentage volume in composite and matrix type. With respect to characterization, the chapter provides an assemblage of knowledge and information for prospective researchers as a one-stop medium toward studies relating to OPF-polymer composite formation.

Section III initiates with *Chapter 5*, which discusses the application of numerical simulations on the biobased adhesive plywood house structure subjected to self-weight and wind loads. Plywood, particleboards and medium density fiberboards are currently used in the house construction, furniture, partitions and others employing harmful formaldehyde as an adhesive material. This chapter deals with a biobased soy meal adhesive for plywood manufacturing. A comparative study has been made by evaluating the flexural strength using a three-point bending, of the commercially available formaldehyde-based glue-laminated plywood. The proposed bio-adhesive has been characterized using field-emission scanning electron microscopy and X-ray diffraction. The optimal plywood manufacturing parameters (viz. temperature and pressure) employing Taguchi's L_9 orthogonal array analysis of variance (ANOVA) was performed to achieve the maximum modulus of rupture. The chapter also used software such as ANSYS and STAAD-Pro to establish the efficacy of the proposed adhesive.

Chapter 6 works with isothermal mold filling simulations for developing liquid composite molded parts. The initial part of this chapter aims at evaluating the range of applicability of ANSYS to perform mold filling simulations. Specialized RTM packages like PAM-RTM is also used to overcome the shortcomings of the standard available packages. Mold filling simulations are initially performed on simple objects. The latter part of the chapter aims at identifying and utilizing appropriate package to perform mold filling simulations and predict the effective injection strategy for industrially important objects that are currently being manufactured by hand layup.

Chapter 7, the last chapter of this section, provides numerical, experimental and simulation study of natural fiber-based composites using injection molding.

Nowadays, the natural fibers have emerged in the composite materials owing to their environment-friendly and mechanical properties. However, as reinforcement, their use in composite material should be well studied, especially during the injection molding. This chapter deals with the study of effect of injection molding on the natural fiber-based composites in terms of length distribution, dispersion and orientation of fibers. Moreover, the focus has been on the defects of this process on the final injected-molding part such as warpage and shrinkage. Finally, the chapter highlighted that the final properties of the injected-molding composite are attributed to some parameters, including aspect ratio of the fibers, fiber concentration, injection parameters, analytical approach for transient response of functionally graded rectangular plates including the higher order shear deformation effects.

Chapter 8, the starting chapter of *Section IV*, elaborates optimization of injection molding process parameters for minimizing volumetric shrinkage and warpage using Moldflow simulation and Taguchi analysis technique. It has been identified that injection molding process parameters have an important influence on product quality. In this chapter, optimal injection molding conditions for minimizing volumetric shrinkage and warpage for rectangular-shaped tensile specimen were analyzed using the Taguchi method and ANOVA technique. A model of rectangular-shaped tensile test specimen was designed and simulated using Moldflow software to imitate the real operating conditions of an injection molding process. These parameters were optimized using ANOVA technique with respect to the volumetric shrinkage and warpage. The results clearly indicated that packing pressure and melt temperature are the most significant parameters to minimize volumetric shrinkage and warpage.

Chapter 9 performs experimental study on laser transmission welding of thermoplastics. Laser transmission welding is now used in a wide range of application areas, including medical devices, automotive components, electrical and electronic devices, packaging, light and displays, household goods and textile industries. This chapter presents a brief overview of the process of laser transmission welding of thermoplastics with a focus on parameters that govern the welding process and the principal phenomena that affect the quality of the joint. Experimental investigations and parametric analyses were carried out to study the effects of parameters on the quality attributes of the laser transmission welding. Finally, the Taguchi quality loss function was used to find the optimum level of control parameters to obtain desired quality attributes.

The last chapter of the book, that is, *Chapter 10*, which is also the only chapter of *Section V*, provides an insight about the environmental aspect of polymers and polymer composites. It has segregated PE, one of the mostly used polymers in both household and industrial applications. The abundance of its usage has led to unwarranted consequences on the environment, including its direct and indirect negative consequences on biological organisms. Although there are various methods to deal with the menace, challenges still prevail. This chapter initially dilated on the structure and properties of PP, which provides its inherent characteristics and subsequently highlights direct and indirect environmental and health costs of plastics in general and PE usage in particular. With regard to public health effects, diseases such as infertility; cardiovascular, nervous and reproductive diseases; sexual immaturation; aggressive behavior; breast cancer; and animal hermaphroditism arising from the presence of endocrine disruptor compounds such as bisphenol A and bisphenol S were also identified. Others include insulin resistance and increased waist circumference, which emanates from ingestion of di-(2-ethylhexyl phthalate). The second objective is to argue that there are opportunities for further research into PP with regard to knowledge enhancement toward PE-natural fiber composite formation and biodegradation as a means of managing the environment. Future researchers, practitioners and industrialists could also concentrate on empirical simulation research toward actual public health consequences of PP on humans. It is expected that research in these areas could help save nations billions of dollars through innovations as well as inexpensive techniques, procedures and processes.

Editors thank God for giving the power to believe in passion and hard work and to pursue dreams; this book could not have been completed without the mercy of the Almighty. The editors also thank all the chapter contributors, reviewers, editorial board members, project development editor and the complete team of publisher Verlag Walter de Gruyter GmbH for their support and availability for work on this editorial book.

This book would be fruitful if future researchers, students, technocrats and people from industrial fraternity use the information provided here to make Mother Earth a better place to live in for the future generations and also for the socioeconomic development of mankind.

Kaushik Kumar
J. Paulo Davim

Contents

Preface —— V

List of Contributors —— XIII

Editors' Biography —— XV

Section I: **Composite and Mold Design**

Sushant P. Mhatugade, Ganesh M. Kakandikar, Omkar K. Kulkarni and V.M. Nandedkar
1 Design, optimization and manufacturing of monocomposite carbon/ epoxy leaf spring having varying cross sections —— 3

Fatima-Zahra Semlali Aouragh Hassani, Wafa Ouarhim, Rachid Bouhfid, Abou el kacem Qaiss
2 Effect of mold design on the fiber orientation in the case of injection molding: experiment and simulation —— 25

Section II: **Characterization and Properties**

Raghu Raja Pandiyan Kuppusamy
3 Characterization of an unsaturated polyester resin for liquid composite molding processes —— 45

Emmanuel Baffour-Awuah, Stephen Akinlabi, and Tien-Chien Jen
4 Properties of oil palm lignocellulose fiber and polymer composite: two-and-a-half decade overview —— 57

Section III: **Simulation and Experimentation**

Tanya Buddi, Swadesh Kumar Singh and B. Nageswara Rao
5 Numerical simulations on the bio-based adhesive plywood house structure subjected to self-weight and wind loads —— 89

Raghu Raja Pandiyan Kuppusamy
6 Isothermal mold filling simulations for developing liquid composite molded parts —— 109

Wafa Ouarhim, Fatima-Zahra Semlali Aouragh Hassani, Rachid Bouhfid, Abou el kacem Qaiss

7 Numerical, experimental and simulation study of natural fiber-based composites on injection molding — 121

Section IV: **Optimization**

Sudeepan Jayapalan

8 Optimization of injection molding process parameters for minimizing volumetric shrinkage and warpage using Moldflow simulation and Taguchi analysis technique — 137

Bappa Acherjee

9 Process overview, experimental study and Taguchi quality loss function analysis of laser transmission welding of thermoplastics — 153

Section V: **Environmental Issues**

Emmanuel Baffour-Awuah, Stephen Akinlabi, Tien-Chien Jen

10 The polyethylene pollution challenge: a review — 169

Index — 187

List of Contributors

Bappa Acherjee
Department of Production Engineering
Birla Institute of Technology
Mesra, Ranchi 835215
India

Stephen Akinlabi
Department of Mechanical & Industrial Engineering Technology
Faculty of Engineering and the Built Environment
University of Johannesburg
P. O. Box 524, Auckland Park 2006
South Africa

Fatima-Zahra Semlali Aouragh Hassani
Moroccan Foundation for Advanced Science Innovation and Research (MAScIR)
Institute of Nanomaterials and Nanotechnology (NANOTECH)
Laboratory of Polymer Processing
Rabat
Morocco

Emmanuel Baffour-Awuah
Department of Mechanical Engineering Science, Faculty of Engineering and the Built Environment
University of Johannesburg
P. O. Box 524, Auckland Park 2006
South Africa

Rachid Bouhfid
Moroccan Foundation for Advanced Science Innovation and Research (MAScIR)
Institute of Nanomaterials and Nanotechnology (NANOTECH)
Laboratory of Polymer Processing
Rabat
Morocco

Tanya Buddi
Department of Mechanical Engineering
Koneru Lakshmaiah Education Foundation
Green Fields
Vaddeswaram, Guntur 522 502
India

Sudeepan Jayapalan
Department of Chemical Engineering
Birla Institute of Technology
Mersa, Ranchi 835215
India

Tien-Chien Jen
Department of Mechanical Engineering Science, Faculty of Engineering and the Built Environment
University of Johannesburg
P. O. Box 524, Auckland Park 2006
South Africa

Ganesh M. Kakandikar
Department of Mechanical Engineering
MIT, Pune
India

Omkar K. Kulkarni
Department of Mechanical Engineering
MIT, Pune
India

Raghu Raja Pandiyan Kuppusamy
Department of Chemical Engineering
National Institute of Technology Warangal
Telangana 506004
India

Sushant P. Mhatugade
Department of Mechanical Engineering
MIT, Pune
India

https://doi.org/10.1515/9783110655049-204

B. Nageswara Rao
Department of Mechanical Engineering
Koneru Lakshmaiah Education Foundation
Green Fields
Vaddeswaram, Guntur 522 502
India

V.M. Nandedkar
Department of Production Engineering
SGGS, Nanded
India

Wafa Ouarhim
Moroccan Foundation for Advanced Science
Innovation and Research (MAScIR)
Institute of Nanomaterials and
Nanotechnology (NANOTECH)
Laboratory of Polymer Processing
Rabat
Morocco

Abou el kacem Qaiss
Moroccan Foundation for Advanced Science
Innovation and Research (MAScIR)
Institute of Nanomaterials and
Nanotechnology (NANOTECH)
Laboratory of Polymer Processing
Rabat
Morocco

Swadesh Kumar Singh
Department of Mechanical Engineering
Gokaraju Rangaraju Institute of Engineering
and Technology (GRIET)
Bachupally, Hyderabad 500 090
India

Editors' Biography

DR. KAUSHIK KUMAR
Associate Professor
Department of Mechanical Engineering
Birla Institute of Technology
Mesra, Ranchi, Jharkhand 835215, India
E-mail: kkumar@bitmesra.ac.in, kaushik.bit@gmail.com

Kaushik Kumar, B.Tech. (Mechanical Engineering, REC (Now NIT), Warangal), MBA (Marketing, IGNOU) and Ph.D. (Engineering, Jadavpur University), is presently an associate professor in the Department of Mechanical Engineering, Birla Institute of Technology, Mesra, Ranchi, India. He has 18 years of teaching and research experience, and over 11 years of industrial experience in a manufacturing unit of global repute. His areas of teaching and research interest are conventional and nonconventional quality management systems, optimization, nonconventional machining, CAD/CAM, rapid prototyping and composites. He has nine patents, 28 books, 19 edited book volumes, 43 book chapters, 141 international journals, 21 international and eight national conference publications to his credit. He is editor in chief, series editor, guest editor, editor, editorial board member and reviewer for international and national journals. He has been felicitated with many awards and honors.

PROF. J. PAULO DAVIM
Professor
Department of Mechanical Engineering
University of Aveiro
Campus Santiago
3810-193 Aveiro
Portugal
E-mail: pdavim@ua.pt

J. Paulo Davim received his Ph.D. in mechanical engineering in 1997, M.Sc. in mechanical engineering (materials and manufacturing processes) in 1991, Mechanical Engineering degree (5 years) in 1986 from the University of Porto (FEUP), the Aggregate title (Full Habilitation) from the University of Coimbra in 2005 and D.Sc. from London Metropolitan University in 2013. He is senior chartered engineer by the Portuguese Institution of Engineers with an MBA and Specialist title in engineering and industrial management. He is also Eur Ing by FEANI-Brussels and Fellow (FIET) by IET-London. Currently, he is professor at the Department of Mechanical Engineering of the University of Aveiro, Portugal. He has more than 30 years of teaching and research experience in manufacturing, materials, mechanical and industrial engineering, with special emphasis in machining and tribology. He has also interest in management, engineering education and higher education for sustainability. He has guided large numbers of postdoc, Ph.D. and master's students as well as coordinated and participated in several financed research projects. He has received several scientific awards. He has worked as an evaluator of projects for European Research Council and other international research agencies as well as examiner of Ph.D. thesis for many universities in different countries. He is the editor in chief of several international journals, guest editor of journals, books editor, book series editor and scientific advisory for many international journals and conferences. At present, he is an editorial board member of 30 international journals and acts

https://doi.org/10.1515/9783110655049-205

as reviewer for more than 100 prestigious Web of Science journals. In addition, he has also published as editor (and coeditor) for more than 100 books and as author (and co-author) for more than 10 books, 80 book chapters and 400 articles in journals and conferences (more than 250 articles in journals indexed in Web of Science core collection/h-index 54+/9000+ citations, SCOPUS/h-index 58+/11500+ citations, Google Scholar/h-index 75+/18500+).

Section I: **Composite and Mold Design**

Sushant P. Mhatugade, Ganesh M. Kakandikar, Omkar K. Kulkarni and V.M. Nandedkar

1 Design, optimization and manufacturing of monocomposite carbon/epoxy leaf spring having varying cross sections

Abstract: The main objective for weight optimization is to replace a steel leaf spring of a light commercial vehicle with a monocomposite leaf spring. Steel leaf spring is modeled in Creo Parametric with the existing dimensions and analyzed for stress and deflection using ANSYS. A composite leaf spring with varying cross sections is designed by using specifications of a steel leaf spring and optimized by using ANSYS. An optimized composite leaf spring is manufactured by the hand layup method with hot molding process. There was 88% weight reduction for composite leaf spring with varying cross sections, which was the main purpose of this chapter. Fabricated leaf spring is tested for the experimental validation. Experimental values were nearly the same as that of analytical with the negligible error.

Keywords: Optimization, composites, leaf spring, ANSYS, manufacturing, monocomposite carbon, epoxy, design, Creo Parametric, nanocomposites, FRP, finite element analysis

1.1 Introduction

As composites, innovation progressed in the course of the most recent couple of decades, the constituent materials, especially the reinforced materials, relentlessly decreased in size. Most as of late, there has been extensive enthusiasm for nanocomposites having nanometer-sized reinforcements, for example, carbon nanoparticles, nanofibers and nanotubes, as a result of the uncommon properties of these materials. There is relentless increment in the significance of polymers, composites and ceramics with the diminishing part of metals. Fibrous reinforcement is extremely powerful in light of the fact that numerous materials are considerably stronger and stiffer in fiber frame than they are in mass shape.

Composite applications in commercial aircraft have been steadily increasing as material costs come down, as design and manufacturing technology evolves and as

Sushant P. Mhatugade, Dr. Ganesh M. Kakandikar, Omkar K. Kulkarni, Department of Mechanical Engineering, MIT, Pune, India
Dr. V.M. Nandedkar, Department of Production Engineering, SGGS, Nanded, India

https://doi.org/10.1515/9783110655049-001

the experience with composites in aircraft continues to build. As an excellent example of innovative design made possible by composites, the use of composites in this airplane resulted in enough weight savings to accommodate the extra weight of an airframe parachute system for safe descent of the entire aircraft in the event of a loss of engine power. The application of composites in commercial airliners has shown steady, conservative growth, but based on the increased prices of fuels, demands by airlines for more efficient aircraft and other recent trends, this growth promises to be rapid in the future.

Focal points of composites over steel are light weight, higher strength-to-weight proportion (up to five times that of steel), no interleaf friction, prevalent fatigue strength, great erosion resistance and higher natural frequency. Composite materials are utilized broadly nowadays in the car business to supplant the metal parts. Springs are crucial suspension components on autos, which are important to limit the vertical vibrations, effects and knocks because of street abnormalities and make an agreeable ride. A leaf spring, notably the longitudinal type, could be a reliable and chronic part in automotive suspension frameworks. These springs are unremarkably formed by stacking leaves of steel, in incessantly longer lengths over one another, with the goal that the spring is thick within the center to oppose twisting and skinny at the finishes wherever it connects to the body. The foremost necessary task is to oppose the variable vertical forces.

Vertical vibrations are supported and absorbed within the spring; therefore, the potential energy is place away in spring as strain energy and subsequently discharged step by step. On these lines, increasing the energy reposition capability of a spring guarantees an additional agreeable suspension framework [1]. The measure of strain energy that can be put away by a leaf spring volume unit is given as follows:

$$S = \frac{1}{2}\frac{\sigma^2}{E} \quad (1.1)$$

where

σ is the maximum allowable stress induced into the spring and E is the modulus of elasticity (both in the longitudinal direction).

1.2 Literature survey

The literature review is carried out to understand and assess the current status. Several papers were dedicated to the application of composite materials for vehicles. Aboul Wafa et al. have completed an investigation of biaxial fatigue of woven roving glass reinforced polyester (GRP) subjected to cyclic bending and torsional moments. Tests are directed to assess failure theories of this material [2]. Adam et al. have made an investigation of the fatigue conduct of a moderate modulus Toray T800 carbon fiber/resin [3]. Sternberg has portrayed the distinctive kinds of suspensions for truck applications [4]. David and Jamie have portrayed the design of fiber reinforced plastic (FRP) composites. They likewise considered the weakness in the design [5]. Arankalle et al. have revealed the advancement of FRP-tapered monoleaf spring for car suspension applications. The models of the composite leaf spring made demonstrate that they are feasible items for light commercial vehicles [6]. Wilson and Bogy have researched the dynamic attributes of a suspension by utilizing a p-type finite element method (FEM). FEM investigation results are confirmed with exploratory outcomes [7]. Shoich Fujiwara et al. have researched the impacts of properties of the material and the glass content on the exhaustion properties of FRP under pulsating tensile stresses [8].

Yu and Kim have examined the prototype longitudinal double-tapered leaf springs to follow that four-leaf steel springs are created using fiber and epoxy and this demonstrated a prevalent perseverance and safeguard attributes [9]. Erol and Mathieu depicted the design and fabrication of a useful composite spring for a sunlight-powered light vehicle. The goal was to give a comprehension of capacities of composite leaf springs created by using unidirectional E-glass wandering fertilized by an epoxy pitch for light-weight vehicle applications wherever the vehicle weight was of essential concern. The plan of application was a sunlight-based fueled auto. ANSYS is utilized for the examination and optimization done using MS Excel. E-glass/epoxy is a material produced by utilizing hand layup strategy [10]. Until now steel leaf springs of automotive suspension are replaced with glass fiber/epoxy material, but there is more deflection than that of steel leaf spring which is undesirable. Hence, there is scope to replace steel leaf spring with carbon fiber/epoxy to reduce excessive deflection. Literature gives stacking sequence optimization as well as geometry optimization of leaf spring having uniform cross section using different algorithms. There is scope to optimize composite leaf spring with varying cross sections using finite element software. This chapter describes the optimization of carbon fiber/epoxy composite leaf spring with varying cross sections by using the finite element tool.

1.3 Design and development of composite leaf spring

1.3.1 Problem statement

To replace the steel leaf spring with a composite leaf spring having varying cross sections used in the rear suspension of a light commercial vehicle and to find out the optimal design of a composite leaf spring that can sustain similar load as that of steel leaf spring by the geometry optimization technique in FEM and its validation with the experimental results.

1.3.2 Objectives

To design and develop the composite leaf spring and to minimize weight and deflection of leaf spring by changing geometrical parameters.

1.3.3 Finite element analysis of steel leaf spring

A three-leaf steel spring of the rear suspension of TATA ACE is selected for replacement with composite leaf spring and its optimization. Parameters of the three-leaf steel spring utilized as a part of this work are as follows.

Material = 65Si7
Span length = 860 mm
Spring width = 60 mm
Spring thickness = 8 mm
No. of leaves = 3
Arc height at axle seat = 90 mm
Mass of half steel spring = 4.86 kg

Load calculation for TATA ACE is given below:

Load of vehicle = 1,200 kg
Max. load carrying capacity = 1,000 kg
Total load = Load of vehicle + max. load carrying capacity
= 2,200 kg × 9.81
= 21,582 N
Load on each leaf spring = total load/number of leaf springs
= 21,582/4
= 5,395.5 N
Load on half leaf = load on each leaf spring/2
= 5,395.5/2
= 2,697.5 N
Factor of safety = 1.5
Load = Load on half leaf × factor of safety
= 4,050 N

1.3.3.1 Modeling of steel leaf spring

This spring is symmetrical so that the length of every half is 430 mm. Each leaf is 60 mm wide and 8 mm thick. A stress analysis was performed utilizing FEM. All the examination was finished by utilizing the version 18.1 of ANSYS. As per the SAE HS 788 standard, a leaf spring can be considered as two cantilevers. Thus, in this chapter rear half of steel leaf spring is analyzed using finite element analysis (FEA). Results for front half are the same as that of rear due to symmetry. A three-dimensional model is created in the Creo Parametric tool as shown in Figure 1.1 and is imported in the ANSYS workbench for further process.

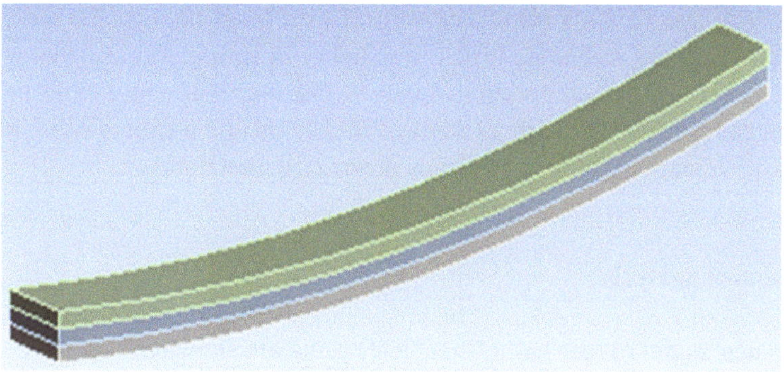

Figure 1.1: Reference model of rear half of steel leaf spring.

Material properties of steel leaf spring are given in Table 1.1. ANSYS enables users to form an acceptable mesh in their analyses through the great toolboxes. There are many distinctive options of ANSYS mesh module like dominant part size, part order and mesh quality. In addition, mesh quality will be verified and increased by means of the mesh module tools in ANSYS. Fine meshing is done by using mesh controls.

Table 1.1: Mechanical properties of 65Si7 [11].

S. no.	Property	Value
1	Modulus of elasticity	200.12 GPa
2	Yield strength	1,081.20 MPa
3	Ultimate tensile strength	1,272 MPa
4	Poisson's ratio	0.266
5	Density	7.85e-6 kg/mm^3

- Element size = 4 mm
- Element type = SOLID186
- Number of elements = 9,990
- Number of nodes = 57,984

Weight of half steel leaf spring is 4.86 kg, so the weight of total leaf spring is 9.72 kg.

1.3.3.2 Boundary conditions

The boundary conditions and also the load to be applied were determined by taking into consideration the higher limit worth of the mechanic masses which will occur, thanks to the vehicle weight. As a result, the vertical load is set because the most dominating and demanding mechanic load is applied on a spring. The calculated total load on each half of the leaf spring is 4,050 N. The rear half of a composite leaf spring is fixed at the axle seat with all degrees of freedom and a load of 4,050 N is applied at the other end (eye end) in vertically downward direction.

1.3.3.3 Finite element analysis

Results for reference model of rear half of steel leaf spring are shown in Figures 1.2 and 1.3. These figures are investigated rigorously as follows. Black outlines show the original shape. Maximum deflection is shown by red color, where no deflection is shown by blue one and all other intermediate deflections are shown by different colors varying from red to blue as shown in figures. From Figure 1.2, it can be observed that deflection is maximum, that is, 76 mm, at eye end, and on the contrary, it is zero at the center of leaf spring. It is nearly 38 mm at the quarter length from the eye end.

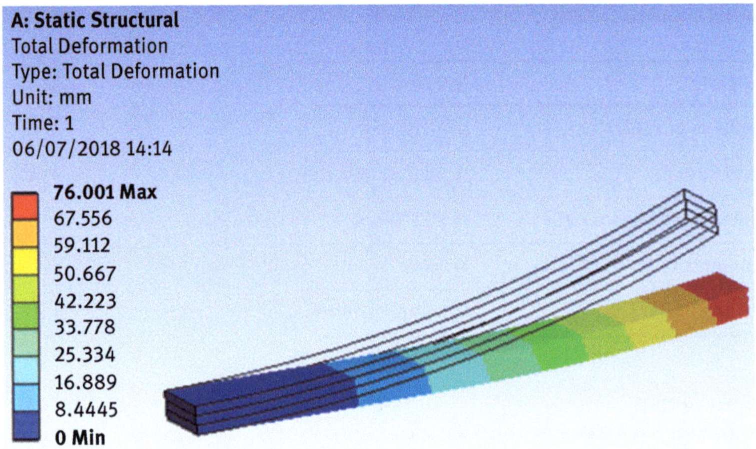

Figure 1.2: Total deformation.

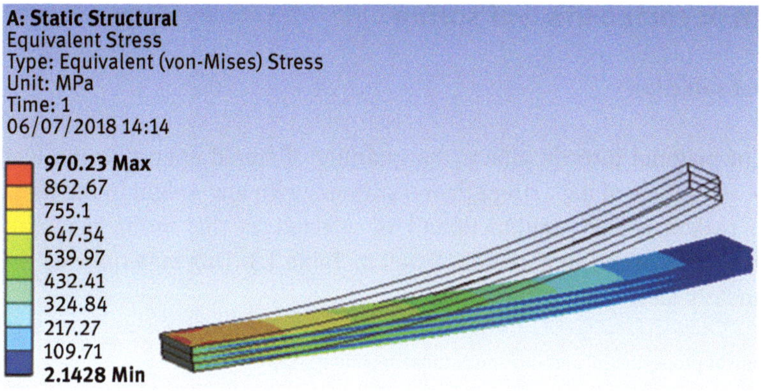

Figure 1..3: Equivalent stress.

a. Total deformation

Maximum equivalent stress is shown by red color and minimum with blue color, where intermediate stress values having different colors are shown in Figure 1.3. From Figure 1.3, it can be observed that the maximum equivalent stress values area are occurring within the middle of the structure on each higher and lower surfaces in terms of the equivalent stresses. It may be discovered from the results that the tensile stresses per unit area dominate more than the compressive stresses on the structure. This could be bestowed as a result of the loading condition. Moreover, from these most stress value areas the strength properties of fabric is less.

b. Equivalent stress

Table 1.2 shows maximum values of output parameters. These results are used for the optimization of monoleaf composite spring, so that it can handle static force similar to steel spring with weight reduction.

Table 1.2: Results (rear half of steel leaf spring).

S. no.	Parameter	Value
1	Maximum total deformation	76 mm
2	Maximum equivalent stress	970.23 MPa

1.3.4 Design of composite leaf spring

1.3.4.1 Material selection

The selection of material directly affects the quantity of stored energy in the leaf spring. To take advantage of high strength and stiffness with low weight in the present work, the carbon/epoxy (unidirectional) is selected as the spring material. Mechanical properties of this material are listed in Table 1.3. This material was assumed to be linearly elastic and orthotropic.

Table 1.3: Mechanical properties of carbon/epoxy (unidirectional) [12].

S. no.	Property	Value
1	Young's modulus *X* direction	121 GPa
2	Young's modulus *Y* direction	8.6 GPa
3	Shear modulus *XY*	8.6 GPa
4	Poisson's ratio *XY*	0.27
5	Density	1,490 kg/m^3
6	Tensile strength *X* direction	2,231 MPa
7	Compressive strength *X* direction	1,082 MPa
8	Tensile strength *Y* direction	29 MPa
9	Compressive strength *Y* direction	100 MPa
10	Shear strength *XY*	80 MPa

1.3.4.2 Layup selection

The layup is chosen to be unidirectional along the longitudinal heading of the spring. The unidirectional layup may debilitate the spring at the mechanical joint territory and require reinforcing the spring in this area.

1.3.5 Optimization of composite leaf spring – varying cross sections

1.3.5.1 Modeling of composite leaf spring with varying cross sections

Thickness is taken as 15 mm at the eye end and 25 mm at the middle end. As we go from the eye end to the center of leaf spring, thickness increases and it is maximum at the center. Width is 55 mm at the eye end and 35 mm at the middle of leaf spring. From the eye end to the middle of leaf spring, width decreases and it is maximum at the eye end.

Other parameters for composite leaf spring are as follows.

Span length = 860 mm
Arc height at axle seat = 90 mm
Normal static loading = 4,050 N

Radius of curvatures responsible for thickness change are taken as 1,072.22 and 974.50 mm, respectively. Radius of curvatures responsible for width change is taken as 9,250 mm as shown in Figure 1.4.

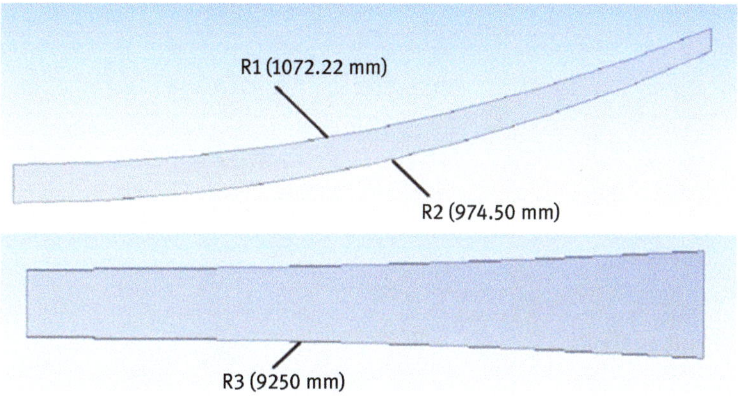

Figure 1.4: Design variables.

According to the SAE HS 788 standards, a leaf spring can be considered as two cantilevers. Therefore, in this chapter, the rear half of spring was optimized as a cantilever beam and results for front half are same as that of rear due to symmetry. Model is created in Creo Parametric tool and imported in ANSYS workbench for further process.

Material properties of composite leaf spring are given in Table 1.3. Meshing is defined as the process of dividing the whole component into the number of elements, so that whenever the load is applied on the component it distributes the load uniformly. In account of the same, mesh quality can be verified and enhanced by way of the mesh tools in ANSYS. Relevance center is set to be fine to obtain better results by using mesh controls. Other mesh data are as follows:
- Element size = 4 mm
- Element type = SOLID186
- Number of elements = 5,550
- Number of nodes = 27,598

1.3.5.2 Boundary conditions

The boundary conditions and the load to be applied were determined by taking into thought the higher limit value of the mechanic masses that will be passed because of the vehicle weight. As a result, the vertical load is set because the most dominating and demanding mechanic load is applied on a leaf spring. The calculated total load on each half of leaf spring is 4,050 N. Loading conditions are same as that of previous case. The corresponding load on each half is 4,050 N. Rear half of composite leaf spring is fixed at one end and load of 4,050 N is applied at the other end, which is the eye end as shown in Figure 1.5.

Figure 1.5: Boundary conditions.

1.3.5.3 Finite element analysis

Composite leaf springs are checked by the leaf spring test rig. Thus, experimental results are obtained from there. FEA could be a convenient methodology for obtaining stress and strain results of the leaf springs. Researchers can compare FEA results with the experimental results to validate their models. Consequently, it is necessary to obtain close results between them. Variable thicknesses and variable widths of semi-elliptical and cantilever composite spring are modeled. The CREO constant quantity was used for modeling, and ANSYS computer code for the analysis.

a. Total deformation

Deformation and stress components that were attained from the FEA were equated with the steel leaf spring's. Figure 1.6 shows results for the reference model of rear half of the composite leaf spring. Using these results by performing the design of experiments (DOE) we will optimize the composite leaf spring for various objectives.

b. Equivalent stress

In this analysis, carbon fiber/epoxy composite leaf spring model was investigated. On the basis of analysis results, the performance and mechanical properties of the

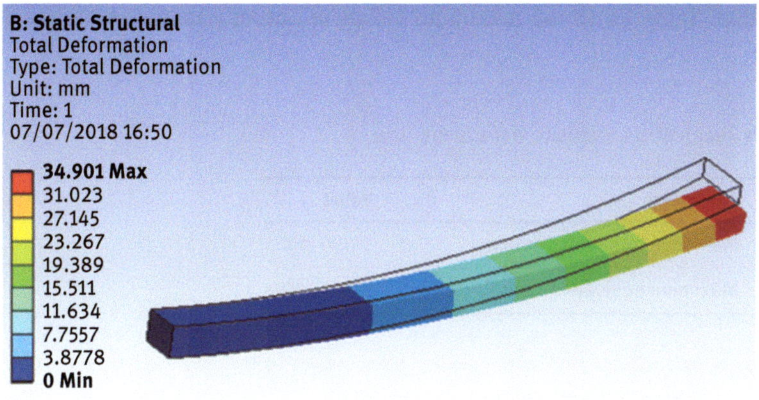

Figure 1.6: Total deformation.

structure in terms of the stress and deflection are designated through Figures 1.6 and 1.7. As shown in the figure, displacement increases linearly as we move from the center to the eye end of composite leaf spring. It shows less deflection than that of steel leaf spring. This situation is qualified to the stiffer nature of the carbon fibers.

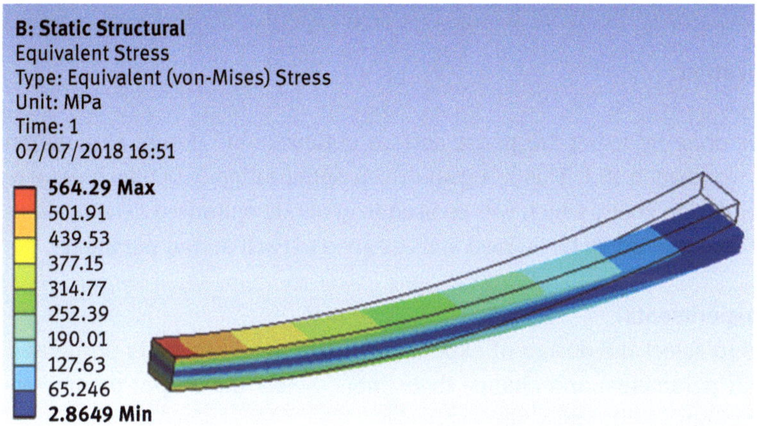

Figure 1.7: Equivalent stress.

Upper surface at the middle of leaf spring having critical region for stress is the tensile one. Figures indicate that the maximum stress values are acceptable again when the strength of carbon/fiber epoxy composite ply is measured and that the stress distributions demonstrate a similar trend with composite leaf spring with constant width and thickness but having lesser deflection values for similar loading condition. Table 1.4 shows maximum values of output parameters obtained by analysis.

Table 1.4: Results (rear half of composite leaf spring).

S. no.	Parameter	Value
1	Maximum total deformation	34.90 mm
2	Maximum equivalent stress	564.29 MPa

1.3.5.4 Design variables, objectives and constraints

Width and thickness of spring are obvious candidates as the design variables but they are varying along length, so that radii of curvature along thickness as well as width are taken as design variables:

$$X = [R_1, R_2, R_3] \tag{1.2}$$

Constraints are dependent variables and functions of the design variables that constrain the design. Stresses and deformations are limited in the spring and must be considered as constraints.

1.3.5.5 Optimization

Optimization is done by using "response surface optimization" design exploration tool of ANSYS workbench 18.1. This is a goal-driven optimization (GDO) technique to state a series of design goals, which will be used to generate optimized designs. Both objectives and constraints can be defined and assigned to each output parameter.

a. Design of experiments

DOE allows us to select the design of experimental cell and change its properties. Select the input parameters and change their limits. Select the output parameters and view their minimum and maximum values.

R_1 is the upper curvature of thickness; R_2 is the lower curvature of thickness; and R_3 the symmetrical curvatures containing width.

Output parameters:

1. Geometry mass
2. Maximum total deformation

By DOE, we obtain the values of output parameters used for optimization. It varies design variables randomly in between upper and lower bounds as shown in

Table 1.5. From this, we obtain the maximum as well as minimum values of output parameters with respect to input design variables.

Table 1.5: Input parameters.

S. o.	Parameter	Lower bound	Upper bound
1	R1	1,062.2 mm	1,082.2 mm
2	R2	964.5 mm	984.5 mm
3	R3	9,240 mm	9,260 mm

All design points evaluated in DOEs are shown in Table 1.6. From DOE, we obtain minimum and maximum output parameters as shown in Table 1.7.

Table 1.6: Design of experiments.

S. no.	R_1 (mm)	R_2 (mm)	R_3 (mm)	Geometry mass (kg)	Maximum total deformation (mm)
1	1,072.2	974.5	9,250	0.6407	34.90
2	1,072.2	964.5	9,250	0.5766	28.23
3	1,072.2	984.5	9,250	0.7047	44.44
4	1,062.2	974.5	9,250	0.5766	42.74
5	1,082.2	974.5	9,250	0.7047	29.27
6	1,072.2	974.5	9,240	0.5189	34.91
7	1,072.2	974.5	9,260	0.6343	34.89
8	1,064.1	966.37	9,241.9	0.6343	33.86
9	1,064.1	982.63	9,241.9	0.7752	50.93

Table 1.7: Output parameters.

S. no.	Output parameter	Minimum	Maximum
1	Geometry mass	0.557 kg	0.584 kg
2	Maximum total deformation	25.74 mm	50.937 mm

b. Optimization

Response surface method is a GDO method, which is a constrained, multiobjective optimization technique in which the "best" possible designs are obtained from a sample set given the objectives or constraints you set for parameters. Response surface optimization contains an optimization component. The DOE and response surface cells are used.

As shown in Table 1.8, we are going to minimize geometry mass, maximum total deformation and maximum equivalent stress.

Table 1.8: Objectives.

S. no.	Name	Parameter	Objective
1	Minimize P4; 30 mm ≤ P3 ≤ 70 mm	P4 – total deformation	Minimize
2	Minimize P5	P5 – geometry mass	Minimize

- **Optimization method used**
  Screening – The screening optimization method uses a simple approach based on sampling and sorting. It supports multiple objectives and constraints as well as all types of input parameters. Usually it is used for preliminary design, which may lead you to apply other methods for more refined optimization results.
- **Configuration –** Generate 1,000 samples and find three candidates.
- **Status –** Converged after 1,000 evaluations.

1.3.5.6 Results

Solution is converged after 1,000 evaluations. Figure 1.8 shows the trade-off chart of total deformation maximum versus geometry mass. Blue color shows feasible points. From this pareto front, three candidate points will be selected within blue points, which will be the optimized one as shown in Table 1.9.

Curvature variation for R_1 is random and changes drastically with respect to iterations. It can be seen that there is a linear variation of optimized curvature of R_2 up to 1,000 iterations. For R_3, it shows the same behavior as that of curvature R_1.

From pareto optimal fronts, three candidate points are selected, which gives the optimum solution as shown in Table 1.9.

By compering these three candidate points, candidate point 2 is having a deflection value greater than the other two, so it will be less stiff than the remaining

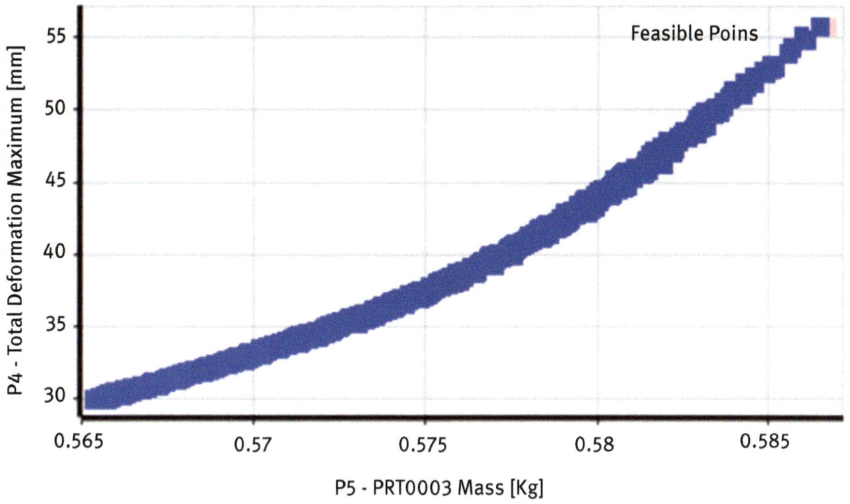

Figure 1.8: Trade-off chart.

Table 1.9: Result of optimization.

S. no.	Parameters	Candidate point 1	Candidate point 2	Candidate point 3
1	R_1	1,062.2 mm	1,064.9 mm	1,070.3 mm
2	R_2	964.51 mm	970.99 mm	974.23 mm
3	R_3	9,240 mm	9,240.1 mm	9,240.1 mm
4	Maximum total deformation	33.633 mm	37.065 mm	35.968 mm
5	Geometry mass	0.5704 kg	0.5743 kg	0.5734 kg

two candidate points. So, we go for candidate point 2, whose optimized parameters are as follows:

R_1 = 1,064.9 mm
R_2 = 970.99 mm
R_3 = 9,240.1 mm

which are the parameters of our optimized geometry of a composite leaf spring. These parameters will be used for the fabrication. Weight of optimized composite leaf is 1.1486 kg.

Weight reduction can be calculated as follows:

$$\text{Weight reduction} = \frac{\text{Wt of steel leaf spring} - \text{Wt of optimized composite leaf spring}}{\text{Wt of steel leaf spring}} \times 100$$

$$= \frac{9.72 - 1.1486}{9.72}$$

$$= 88.18\%$$

1.4 Manufacturing of optimized leaf spring

With the hand layup procedure, we can produce monocomposite leaf spring. For well-being reason, hand claps and veils are utilized, since it is hazardous to well-being.

1.4.1 Mold manufacturing

Initially, the scale of optimized composite leaf spring (light car rear leaf spring) is noted in keeping with the scale the mold is built by using sheet metal of thickness 5 mm. Upper and lower plates are rolled to get desired camber height 90 mm. Side plates are cut down using profile cutting by laser technology. These side plates are welded to bottom plate as shown in Figure 1.9.

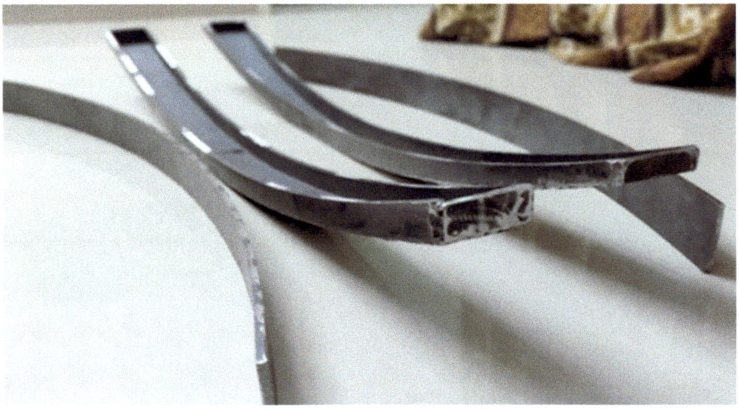

Figure 1.9: Fabricated mold design.

For easy release of composite leaf spring from the mold, C-type channels are made by welding three plates together as shown in Figure 1.10. These C-type plates are attached at both the eye ends using bolts.

Figure 1.10: C-type channel.

Buffing of mold is done to get better surface finish. The prepared mold is heated and refined with wax (which acts as a freeing agent) in order to exclude sticking of the composite resin after preserving.

1.4.2 Preparation for molding

The total amount of carbon fiber cloth required to construct the composite leaf spring is calculated based on the scale of the optimized composite spring and also the thickness of every plate. To avoid delamination, one layer of bidirectional carbon fiber is used after two unidirectional layers. Then the carbon fiber cloth is weighed on the digital weighing machine. This cloth is then cut down with the required dimension (curved profiles for varying *W*, *T*). Corresponding to the weight of carbon fiber, epoxy resin is taken at a ratio of 70:30 and 10% hardener with accelerator is added to it. Resin hardener mixture is mixed systematically and stirred till a gel-like look appeared.

1.4.3 Hand layup procedure

After the application of a release agent, mold is heated in an oven for some time and the resin is applied on the heated mold. The carbon fiber cloth is placed in the mold and resin is applied on the carbon fiber without distorting the fibers with the help of a brush, and any air stuck between the cloth and the surface is removed by stamping using a roller.

The soaking is done equally by stirring a hand roller on the cloth surface so that the fiber should not be isolated from the cloth. Simultaneous laying of laminas on the die is done to the essential thickness of the laminate; 25 layers of unidirectional carbon fabric (t = 0.6 mm) and 12 layers of bidirectional carbon fabric (t = 0.22 mm) are layered to fill the mold.

After the molding process, the top curved plate is placed and clamped with C-clamps at different places. For the curing process, this set is placed in an oven at 120 °C for 90 min. After removing from the oven, the composite leaf spring is released by removing C-channels bolted at both ends.

1.4.4 Post molding process

In the molding process, both pressures, that is, positive and negative, have been applied on the laminate in the form of C-clamping as well as heating in the oven. Therefore, excessive resin will get out of the mold and cure there itself on the upper edges of the leaf spring.

Excessive material is then removed by the grinding process. Hand grinder is used for this removal process. Sharp edges are removed by using a file. After the grinding process, the composite leaf spring is polished with the help of sand paper of grades 80 and 120, respectively.

Composite leaf spring with varying W, T is fabricated by the same way as described earlier. This one is polished by the same way but painted by applying varnish on it and heated in an oven for some time. For better surface finish, leaf springs are polished with a sand paper in water. Finally, Vaseline is applied to obtain the shiny surface.

1.4.5 Eye joints

A sheet metal piece having 5 mm thickness and required width is rolled to make U shape out of it. The pipe of mild steel having outer diameter equal to inner semicircular diameter of U shape is welded at the inner side of the semicircular shape as shown in Figure 1.11. These eye joints are bolted to both composite leaf springs with three nut-bolts at each end with the help of a drill.

1.5 Experimental validation

1.5.1 Experimental setup

Both the leaf springs are tested on UTM (universal testing machine) to know whether they can withstand at designed load and give deflection values as FEA. In the

Figure 1.11: Composite leaf spring with eye joint (varying *W*, *T*).

optimization process, half-leaf spring is considered. Load applied was 4,050 N. Here we are using complete leaf spring so that load to be applied will be double of it, that is, 8,100 N.

C-channel beam is used as a fixture for testing of composite leaf spring as shown in Figure 1.12. This fixture is bolted on the bed of UTM. Composite leaf spring is placed upside down on that fixture. One end of the leaf spring is fixed by applying support and the other end is free to move along the longitudinal axis of a composite leaf spring as shown in Figure 1.12. Load is applied gradually up to 9,000 N at the center of leaf spring and deflection of leaf spring is recorded as given in Table 1.10.

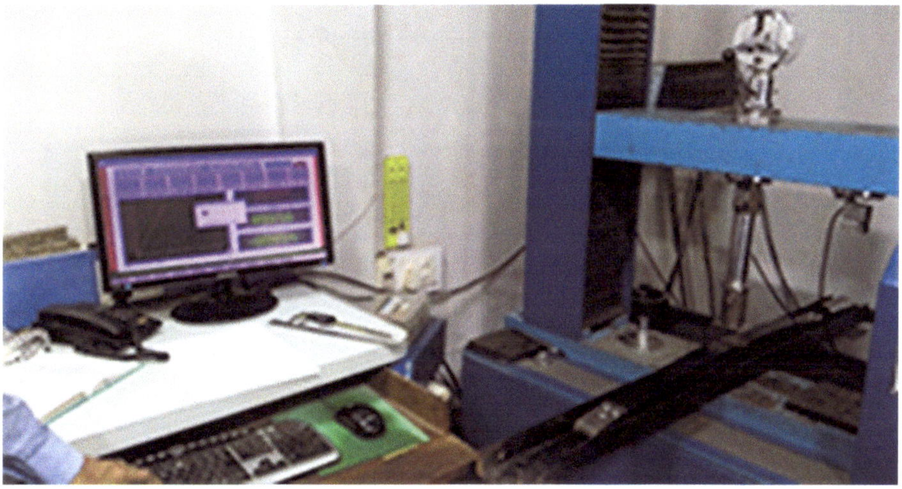

Figure 1.12: Testing of composite leaf spring on UTM.

Table 1.10: Load and deflection results for composite leaf spring (varying *W*, *T*).

S. no.	Time (s)	Load (N)	Elongation (mm)
1	0.5	42.1	0.1
2	45.7	2,990	15.4
3	70.2	5,016.6	23.7
4	106.9	8,103.6	36.1
5	120.2	9,060.1	40.6

1.5.2 Experimental results

Table 1.10 shows the variation of elongation of composite leaf spring with respect to load as well as time. Table is formulated by picking up five instances in between.

Load versus deflection graph for composite leaf spring with varying cross sections is shown in Figure 1.13. From the graph, it can be noted that deflection is varying linearly with the applied load. At full load of 9,060 N, it shows 40.6 mm where for the designed load of 8,100 N, it is 36 mm, which is quite near to the FEA results.

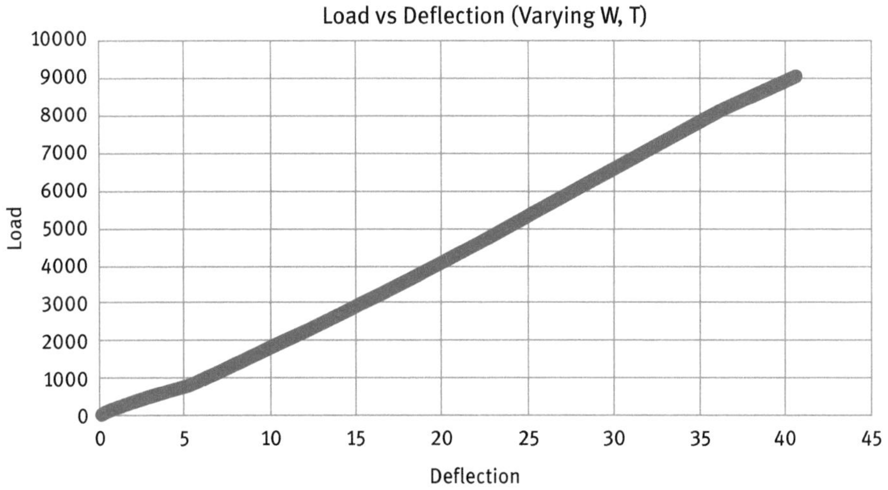

Figure 1.13: Load versus deflection (varying *W*, *T*).

For designed load, error between FEA results and experimental results is calculated as follows:

For constant W, T, composite leaf spring

$$\text{Error} = \frac{\text{Experimental deflection} - \text{deflection by FEA}}{\text{Experimental deflection}} \times 100$$

$$= \frac{36.1 - 37.065}{36.1}$$

$$= -2.67\%$$

From this, it can be concluded that FEA and experimental results are nearly the same.

1.5.3 Concluding remarks and scope for the future work

After replacing the steel leaf spring with the composite, there is weight reduction of nearly 88% for composite leaf spring having varying widths and thicknesses. Experimental results are in line with the obtained result with the small error of 2.67%.

The two methods used for optimization are FEA and Salp Swarm Algorithm. It can be optimized by using different methods by taking different composite materials with the combination of varying widths and thicknesses. Fatigue life can be increased by optimizing the same. Different methods other than the hand layup like resin transfer molding and vacuum bagging process can be used for the fabrication of composite leaf spring.

References

[1] Shokrieh, Mahmood M. Rezaei, Davood. "Analysis and optimization of a composite leaf spring". Compos Struct 2003, 60(3), 317–325.

[2] Wafa, MNAboul., Hamdy, A. H., and El-Midany., A. A. "Combined bending and torsional fatigue of woven roving GRP". J Eng Mater Technol 1997, 119(2), 180–185.

[3] Harris, B. et al. "Fatigue behaviour of carbon fibre reinforced plastics". Composites 1990, 21(3), 232–242.

[4] Sternberg, Ernest R. Heavy-duty truck suspensions. SAE Trans 1976, 1345–1396. https://doi.org/10.4271/760369.

[5] Riegner, David A., and Hsu, Jamie C. Fatigue considerations for FRP composites. SAE Technical Paper 1982, 820698.

[6] Bajpai, Ram P., Chandrasekhar, U., and Arankalle, Avinash R. eds. Innovative Design, Analysis and Development Practices in Aerospace and Automotive Engineering: I-DAD, 2014, February 22–24, 2014, Springer Science & Business, 2014.

[7] Wilson, C. J., and Bogy, D. B. "Modal analysis of a suspension assembly". J Eng Ind 1994, 116(3), 377–386.

[8] Fujiwara, Shoichi. et al. "Tensile Fatigue Properties of FRP". Bull JSME 1973, 16(101), 1648–1656.

[9] Yu, WJ., and Kim, HC. Double tapered FRP beam for automotive- suspension leaf spring. Comp Struct 1988, 9, 279–300.
[10] Erol, Sancaktar., and Mathieu, Gratton. Design, analysis, and optimization of composite leaf springs for light vehicle applications. Comp Struct 1999, 44, 195–204.
[11] Arora, Vinkel Kumar., Gian, Bhushan., and Aggarwal, M. L. Fatigue life assessment of 65si7 leaf springs: A comparative study. International scholarly research notices 2014, (2014).
[12] ANSYS Workbench Reference. Release 18.1. ANSYS, Inc. 2018.

Fatima-Zahra Semlali Aouragh Hassani, Wafa Ouarhim,
Rachid Bouhfid, Abou el kacem Qaiss

2 Effect of mold design on the fiber orientation in the case of injection molding: experiment and simulation

Abstract: Injection molding represents one of the most important processes to manufacture plastic parts, especially for complex geometries. However, injection of short fiber-reinforced thermoplastics along with their inherent processing advantages shows high-composite performance (high strength-to-weight ratio, high stiffness and possibly controlled anisotropy). Indeed, flow of a fiber suspension during injection molding results in a short-fiber preferential orientation along the mold width and thickness. In this chapter, parameters affecting the fiber orientation such as the mold design (especially cavity shape and gate position) during the injection molding were detailed. Then, to predict the fiber orientation during the shaping phase, various digital software able to solve the problems related to the injection of composite such as REM 3D software and Moldflow insight were presented. Finally, since the mechanical properties of a composite are largely affected by the fiber properties, several models of the composite mechanical performance prediction were depicted.

Keywords: Injection molding, Mold design, fiber orientation, Simulation.

2.1 Introduction

Nowadays, composite materials have been extensively used for their important role in attenuation of components design, low specific weight, high mechanical performance and excellent corrosion resistance, which offer significant advantages over metallic materials [1, 2]. As the objective is to increase both the productivity and the mechanical quality of parts, the composite materials are obtained by an injection process of the polymer matrix and its reinforcement. This method makes it possible to obtain parts of various geometric sizes and complexities and remains an economical process, well suited to large series, giving dimensional tolerances generally sufficient to overcome machining operations. These composite materials thus have the advantage of being

Fatima-Zahra Semlali Aouragh Hassani, Wafa Ouarhim, Moroccan Foundation for Advanced Science, Innovation and Research (MAScIR), Composites and Nanocomposites Center, Rabat, Morocco; Mohammed V-Rabat, University, Faculty of Science, Mechanics of Materials Laboratory, Rabat, Morocco
Rachid Bouhfid, Abou el kacem Qaiss, Moroccan Foundation for Advanced Science, Innovation and Research (MAScIR), Composites and Nanocomposites Center, Rabat, Morocco

https://doi.org/10.1515/9783110655049-002

able to be shaped by the same processes as the uncharged thermoplastics and this without much modification of the tools [2–4].

However, during the flow phase of the charge in the mold, each fiber is transported by the flow and its orientation evolves according to the constraints imposed by the matrix, the other particles and the walls of the mold. This results in a complex orientation distribution, which varies considerably in the part, in particular according to the thickness, thus affecting the thermomechanical properties such as the elastic modulus, the tensile stress, the impact resistance as well as the coefficient of expansion. This will depend not only on this orientation distribution but also on the local concentration of the fibers as well as the heterogeneity of the fiber length distribution [5, 6].

Anisotropy, induced by the presence of fibers, is an important property that must be taken into account when designing parts and mold. Indeed, the orientation of the fibers has a major role on the strength and rigidity of a room. At the same time, the heterogeneity induced by a length distribution of the fibers and a nonhomogeneous concentration in the flow is often a source of defects that can lead to a warping of the parts that are noticeably predictable or that can induce an early fatigue of the composite material [6, 7].

It therefore seems essential to control or at least predict the orientation of the fibers during the shaping phase. Therefore, this chapter aims at providing a state of the art regarding fiber orientation in composite injection molding, summarizing the main results of recent research. So, after a general description of the process of injection molding of filled thermoplastics, the first part of this chapter was devoted to the qualitative analysis of the orientation mechanisms encountered during the shaping process as well as various parameters influencing it. The second part was dedicated to the presentation of various numerical simulation tools, given its ability to predict the flow of the material and therefore the orientation of the fibers in the mold. Finally, we finish by macroscopic modeling of the fiber orientation with in particular the description of the mathematical tools used during the modeling of the fiber orientation.

2.2 Fiber orientation in the case of injection molding

2.2.1 Injection molding

Injection molding represents one of the most important and popular molding technologies to manufacture plastic parts, especially those with complex geometry due to its low cost, high productivity and high precision [4, 8]. A typical injection molding cycle involves four distinct stages, namely clamping, injection, cooling and ejection stages (Figure 2.1) [4, 8]. At first, the granules fed into the injection molding machine are gradually heated, melted and homogenized by the shear due to the rotation of the screw. The energy required for fusion comes from the combined action of mechanical dissipation and heat transfer from the heated sheath. This step is one of the origins of the decrease in the average length of the fibers, and will thus induce a nonhomogeneous length distribution in the flow. At the chosen moment, the screw is pushed axially and plays the role of a piston by injecting at a controlled speed the molten polymer in the cavity of the mold. During filling, the fibers will be oriented in preferred directions according to the stresses involved. Once mold cavities have been filled, high pressure is maintained as far as possible until the material is frozen. An additional quantity of polymer is then introduced into the cavity, in order to limit the heat shrinkage and gradually standardize the pressure in the mold. Finally, the piece remains a few moments in the mold so that cooling can continue. Once the part is cooled, the mold opens and a mechanism to push the part out of the mold force is applied to eject the part because during cooling the part shrinks and adheres to the mold. During this time, the clamping phase starts again, so as to prepare the next cycle [4, 8].

This results in a complex orientation distribution, varying considerably in the room, in particular according to the thickness. The thermomechanical properties will depend on this orientation distribution but also on the local concentration of fibers as well as the heterogeneity of the fiber length distribution. The mathematical and numerical (simulation) tools necessary for the prediction and control of this orientation will be presented in the following sections.

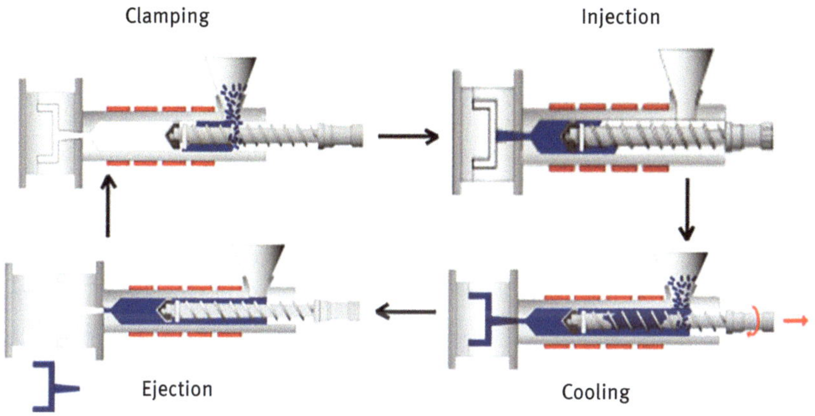

Figure 2.1: Injection molding process cycle.

2.2.2 Injection orientation mechanisms

During the filling phase, the fibers will orient themselves in privileged directions, creating a structure that is commonly called core–skin. In this section, a brief state of the art of knowledge in this domain was proposed.

Two types of flow will preferentially orient the fibers:

Shear flow: Such flow occurs, for example, in the mold supply channels and/ or between parallel plates. A fiber isolated in a Newtonian fluid is animated by a periodic rotation movement, with a privileged residence time in the direction of the flow: the fiber never tends toward a position of equilibrium [9, 10].
Elongational flow: We meet it, for example, on the axis of geometry convergent or divergent. The observation of a fiber in this type of geometry shows that the fiber is oriented in the divergent perpendicular to the direction of flow and converges parallel to this direction. The fiber tends to a position of stable equilibrium perpendicular or parallel to the flow [9, 10].

In the following paragraphs, we will review the various factors that are likely to affect the fiber orientation.

2.3 Influence of the main injection parameters on the fiber orientation

2.3.1 Influence of mold and melt temperature

It is well known that decreasing mold or melt temperature lead necessary to the rapid solidification of the polymer, thus inducing maximum shear values shifted toward the center of the part. This decrease in the temperature results in an increase in the shell–skin layer thickness. The phenomena observed are mainly related to the fountain effect, which is accentuated by the cooling rate of the polymer [11]. Bay et al. [12] found different results, which they nicknamed fortunate when evaluating the effect of a slight variation of mold and melt temperature on fiber orientation for a center-gated disk. So, when comparing the fiber orientation of three mold sets successively at T_w = 337, 347 and 357 K, no significant influence was observed, but the molecular orientation or crystallinity in some polymers may greatly be affected. On the other hand, the same comparison at three melt inlet temperature values T_{in} = 540, 550, and 560 K depicts that fiber orientations are not much sensitive to T_{in}, apart from the fact of adjusting the transition layer between the core and shell layers.

2.3.2 Influence of filling time and injection speed

Generally, the core layer increase and the skin layer diminution are highly caused by high injection speed. Indeed, the fountain flow greatly affects the skin area by not allowing the fibers time to orient them, thus shifting the high shear region toward the mold walls [11]. Bay and Tucker [12] investigate the effect of the filling time on fiber orientation by imposing four filling time values t_{fill} = 0.5, 2, 2.5 and

3 s. Results emphasize that the filing speed significantly control the fiber orientation and the thickness of different layer. Indeed, for a very fast filling rate, there is virtually no skin layer [12]. While Chung and Kwon [13] evaluated the core fiber orientation at different filling times (t = 0.081, 0.255, 0.557 and 0.81 s) for a tensile testing specimen. Results show that by increasing the filling time the fibers tend to align in the flow direction [13].

2.4 Influence of the main material parameters on the fiber orientation

Three essential parameters can influence fiber orientation in an injection molded part: matrix rheology, fiber concentration and fiber aspect ratio. We will review these parameters and in particular their impact on fiber orientation.

2.4.1 Influence of the matrix

The existence of a flat velocity profile in the cavity is the result of the pseudoplastic character of the polymers. The rheology of the material, at high shear rates, influences the shape of the velocity profile and therefore its ability to convey the orientation acquired in the threshold in the rest of the cavity. On the other hand, it is also shown that the convection of a transversely oriented core in the cavity without reorientation is facilitated by the presence of the fibers, which can flatten the velocity profile, even in isothermal conditions. It should be kept in mind, however, that many of these effects act in the same way. It should also be noted that, in the vast majority of cases, these studies remain qualitative [11].

To provide a better understanding of the material rheological properties effect on the fiber orientation, Bay and Tucker [12] simulated the filling of the disk using three polymers with different properties (nylon, polypropylene (PP) and polycarbonate (PC)) to conclude that their different heat of transfer and viscosity lead to different orientation distribution. The flatter velocity profile of PP produces a much thicker core layer and shearing near the wall. While the large heat of fusion erases the fountain flow effects, no skin layer is formed. Finally, the PC has the same skin layer as nylon, but a core region smaller than even nylon and PP moldings.

2.4.2 Influence of the reinforcement

An extensive study carried out on fiber-reinforced polymer matrix (short and long) with several reinforcement rates (in weight) shows that the width of the core increases with the fiber concentration. It also appears that the phenomenon is more important in the case of long fibers [14].

Generally, there are three regimes of concentration: the regime where there is no interaction between the fibers is also called dilute solution, the suspension where the fiber distance is between the fiber length and diameter is semiconcentrated solution, and finally, the concentrated solution when the average distance between fibers is lower than the fiber diameter [15].

2.5 Effect of the mold design on the fiber orientation: experimental study

2.5.1 Gate position

The presence of gates leads to a core layer formation with a transverse fiber orientation, when they are the seat of elongation. Conversely, in the case of gate film or carrot injections, a formation of a core layer occurs with a fiber orientation in the flow direction [16].

As an example, Shokri and Bhatnagar [17] tested several asymmetrical gate locations to show the effect of gate location on fiber orientation to conclude that the asymmetric gate position causes an asymmetric viscosity state in the transverse flow direction (width of the mold) and therefore leads to asymmetric fiber patterns in the flow plane.

2.5.2 Mold design

Typically thermoplastic fiber-reinforced polymers have a very particular structure in the thickness that is commonly known as the “core–skin” structure.

The differences between the authors come from the count of the number of layers in the thickness. Kenig [18] depicts a nine-layer structure, with a very thin skin layer (poor quantity of fiber), a sequence of three layers with alternating fibers oriented parallel/transversely/parallel to the main flow direction and a core with a transversal fiber orientation. On the other hand, Bay and Tucker emphasize the presence of seven distinct layers symmetrically with respect to the median plane of a disk injected by the center (Figure 2.2). The two extreme layers (A), called skin layers, have an isotropic orientation of the particles. Then appear two layers (B) where the fibers are oriented in the flow direction. In this area of the disk, shear stresses are important during the flow. The particles are arranged in concentric circles in the central layer (D) (core layer). Here it is the elongation constraints that are predominant during the flow. Finally, the last two layers (C), in which the orientation of the fibers is random, are transition layers between the orientation states (B) and (D) [10, 19].

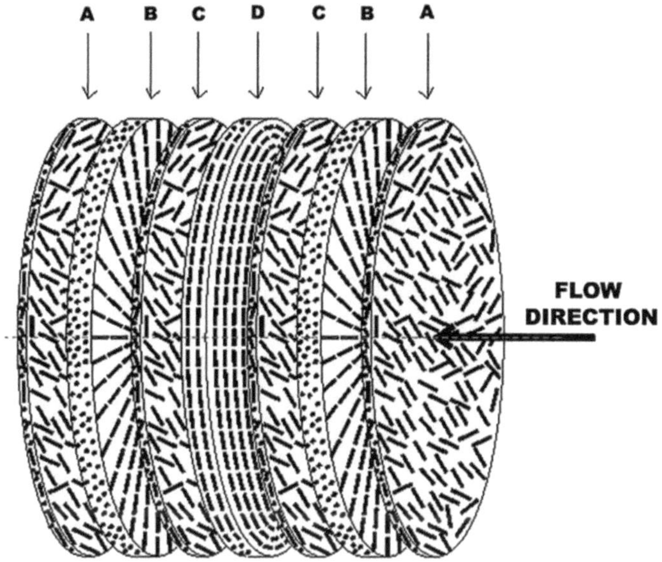

Figure 2.2: Fiber orientation distribution in a disk injected by the center.

Then the two authors no longer speak of transition layers but of a large area where the fibers are gradually oriented perpendicular to the direction of flow.

In the vast majority of cases, under conventional injection conditions, it is simply a five-layer "skin–core" structure with a central layer oriented transversely to the direction of flow, two intermediate layers oriented mainly in the flow direction and finally two skin layers slightly less oriented than the intermediate layers [19].

The formation of a skin layer in which the fibers are not oriented preferentially is the result of the fountain effect (Figure 2.3). The latter returns the material (fibers included) from the center to the cold walls of the mold where the fibers will be frozen with a disordered orientation before the shear has had time to orient them in the direction of flow. Often, because of the very short injection times and the low conductivity of the polymers, this layer is very thin and difficult to identify by observation with

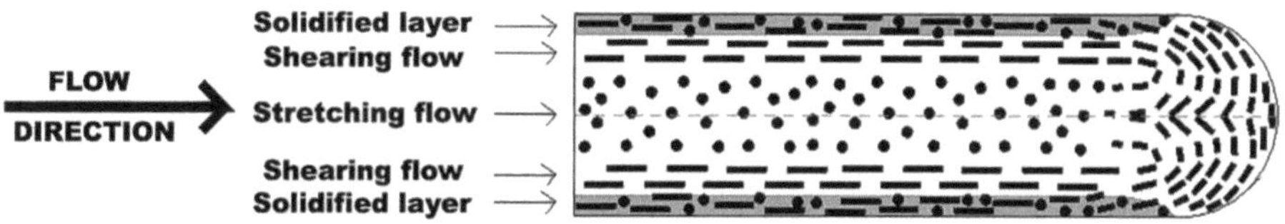

Figure 2.3: Fountain flow effect.

respect to the other zones. The thickness of this layer depends on the cooling rate (mold temperature) [12, 20, 21].

The formation of the intermediate zone, in which the fibers are strongly oriented in the main flow direction, is the consequence of the important shear which develops there [12, 20, 21].

The formation of the core layer, in which the fibers are oriented transversely to the flow direction, comes from the combination of two factors [12, 20, 21]:

- The presence of an elongational flow around the injection threshold, which will align the fibers perpendicularly to the flow direction (this orientation is, moreover, completely independent of the initial orientation of the fibers in the feed channels).
- The conservation, in the center, of this transverse orientation all along the cavity due to practically zero shears. The existence of a flat velocity profile in the cavity, for example, will help to preserve the transverse fibers orientation in the rest of the cavity without creating reorientation. If either of these conditions is not met, the core fibers will be more or less aligned in the direction of flow.

Generally, it has been proved for rectangular mold with uniform thickness that decreasing the cavity thickness decreases the core zone until it disappears, resulting in nearly the same planar orientation throughout all the thickness [11, 16]. Indeed, by comparing the fiber orientation at different mold thicknesses (1.1, 1.7, 3 and 5 mm), Vincent et al. [22] have found that decreasing the mold thickness decreases the core–skin structure until obtaining one layer, where the fibers adopted a specific planar orientation throughout all the thickness. As the plastic contacts the mold surfaces, a frozen layer forms immediately along the mold walls where the converging flow tends to flow-align the fibers, while near the centerline due to decreasing the solid layer, the diverging flow tends to align fibers perpendicular to the flow, hence forming a parabolic shape of fiber orientation [23].

On the other hand, decreasing the mold width also plays a major role in the fiber orientation distribution. Vincent et al. [5] compare the fiber orientation in two different mold widths (20 and 40 mm). To conclude that reducing the mold width leads fibers to orient in the flow direction in the whole thickness, whatever the position taken along the plate.

2.6 Numerical study of the fiber orientation during injection molding

This part deals with the numerical methods of the fiber movement with the fluid simulation to obtain the effects of the different variables influencing the orientation of the fibers.

The principal hypotheses on which the calculation of orientation is based assume that this orientation develops during the flow of the polymer. At the end of the filling, the orientation is supposed to be frozen.

The general framework for numeric resolution of the orientation is the REM3D software. This industrial software solves the problem of injection and extrusion for a generalized Newtonian fluid. In this way, we were able to benefit from previous developments: the mechanical solver, the thermal solver, taking into account of the dependence of the viscosity as a function of temperature and shearing as well as displacement of the interface.

Using the Rem3D software, the logic resolution is to establish a resolution scheme that will execute all Δt since the different numerical variables are time dependent. Indeed, during the flow, the time interval is divided into submeasurement intervals Δt (Δt being the time step) (Figure 2.4).

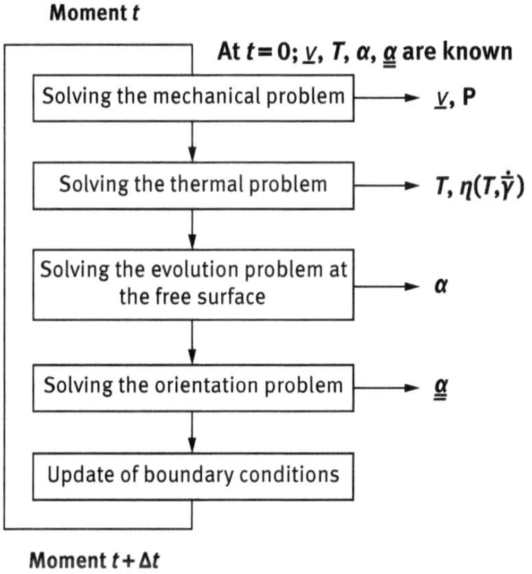

Figure 2.4: Rem3D digital resolution scheme at each increment of time.

The flowchart of Figure 2.4 shows that at a time t, the numerical scheme begins with the resolution of the mechanical problem in order to know the distribution of local velocity $\underline{v}$ in the whole domain; the method of resolution is the method of mixed finite elements in velocity–pressure with interpolation P1 +/P1. The second step of this approach is to solve the thermal problem which is in the form of convection–diffusion equation; the resolution of this equation requires the knowledge of the velocity field and makes it possible to calculate the thermodependent viscosity starting from rheological laws (of the Carreau-Yasuda type, Cross-WLF, etc.). These rheological laws generally depend, in addition to temperature, on the rate of deformation, and consequently on the velocity field, which makes the mechanical problem nonlinear. The third step is to calculate a function α, called distance function, which detects the evolution of the free surface of the molten polymer; this distance function is calculated using a Level Set method. At the end, the orientation tensor is calculated, and the same numerical scheme is repeated at time $t + \Delta t$ [11].

Abla et al. demonstrate that it is currently possible to calculate the fiber orientation in the case of an extrusion or injection process using Rem 3d. First, they presented an example concerning a case of extrusion of a rectangular plate 2d of dimension 5 out of 2. The considered flow is a flow of Poiseuille constant flow type. They have initiated several simulations involving Np values (defined as increasing function of aspect ratio and fiber volume fraction), which may mean increasing the fiber volume fraction. Results showed that the presence of fibers disturbs the velocity profile as if the fluid was "pseudo-plastic": the velocity profile flattens. At the same time, the coupling reveals a central zone, where the fibers are partially oriented in a direction perpendicular to the direction of flow [11].

In the second place, a filling of a 3d piece with the evolution of the free surface has been presented. The rheological parameter Np is zero in this calculation (no coupling). It has been observed that [11]:

- the fibers are oriented mainly in the flow direction close to the walls, or the shear reaches its maximum value;
- the fibers are oriented perpendicular to the flow near the free surface (fountain effect) and
- the fibers at the surfaces of resource are oriented parallel to this surface, which explains the known mechanical weakness of these areas.

Other studies use Autodesk Simulation Moldflow Insight as a predictive tool to study the fiber orientation and length distribution following flowchart presented in Figure 2.5.

Three-dimensional fiber orientations are calculated with the mold filling analysis on the same finite element mesh, and each triangular element contains several layers in the local molding thickness.

The three-dimensional orientation solution for each element is described by a second-order tensor:

$$a_{ij} = \begin{pmatrix} a_{11} & a_{12} & a_{13} \\ a_{21} & a_{22} & a_{23} \\ a_{31} & a_{32} & a_{33} \end{pmatrix} Y \begin{pmatrix} \lambda_1 & 0 & 0 \\ 0 & \lambda_2 & 0 \\ 0 & 0 & \lambda_3 \end{pmatrix}; \; [e_1 \, e_2 \, e_3] \tag{2.1}$$

where a_{11} represents fiber orientation in the flow direction, varying from 0 to 1; a_{22} represents fiber orientation transverse to flow, varying from 0 to 1.0 and a_{13} the tilt of orientation in the 1–3 plane, varying from −0.5 to 0.5.

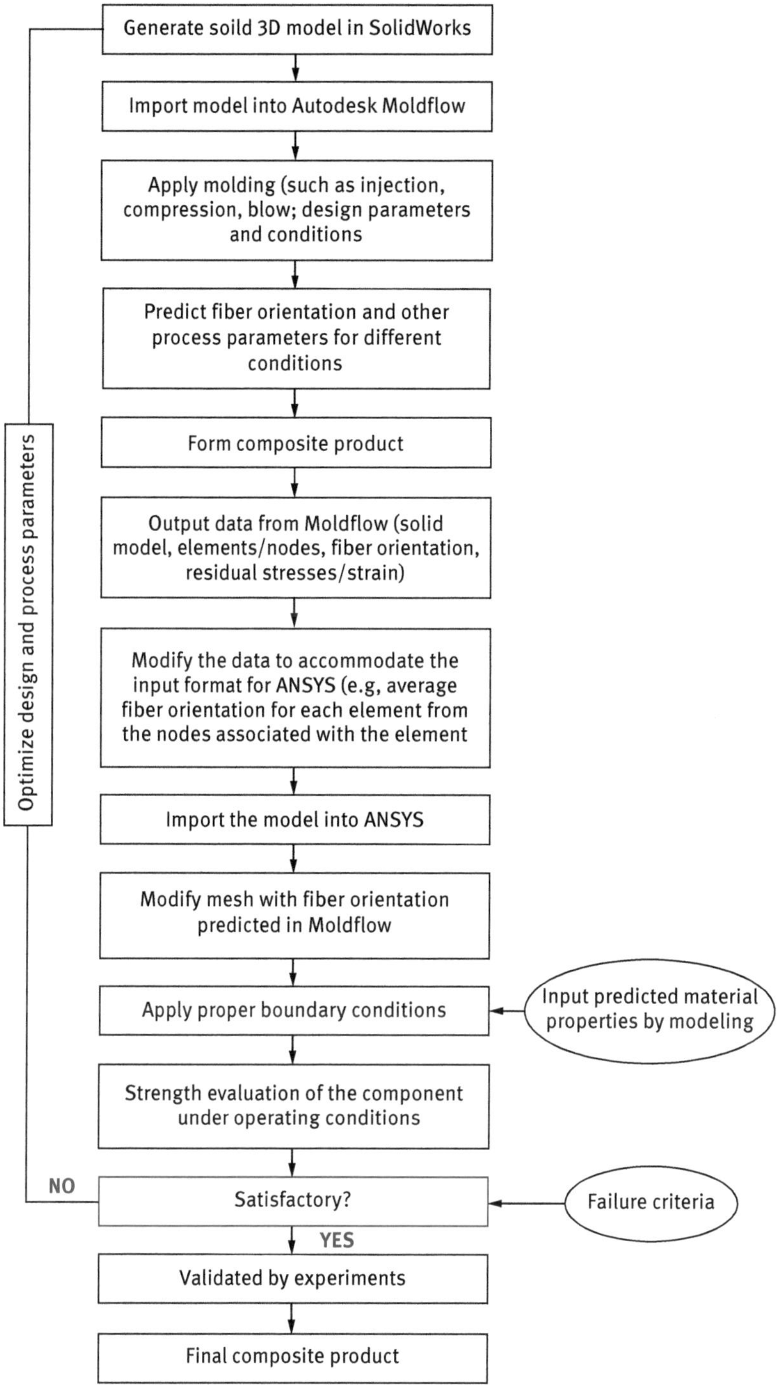

Figure 2.5: Steps used for Autodesk Moldflow Insight Simulation.

2.6.1 Moldflow's fiber orientation models

The equation governing the fiber orientation of Moldflow for 3D meshes is called the Folgar–Tucker orientation equation (eq. (2.2)) [24]:

$$\frac{Da_{ij}}{Dt} = -\frac{1}{2}[w_{ik}\,a_{kj} - a_{ik}\,w_{kj}] + \frac{1}{2}\lambda[\dot{\gamma}_{ik}\,a_{kj} + a_{ik}\,\dot{\gamma}_{kj} - 2a_{ijkl\dot{\gamma}_{kl}} + 2C_{l\dot{\gamma}}(\delta_{ij} - 3a_{ij})] \tag{2.2}$$

where a_{ij} is the fiber orientation tensor, $\frac{1}{2}w_{ij}$ is the vorticity tensor, $\frac{1}{2}\dot{\gamma}_{ij}$ is the deformation rate tensor and C_l is the fiber interaction coefficient, a scalar phenomenological parameter, the value of which is determined by fitting to experimental results.

Governing equations for Midplane and Dual Domain meshes is based on the following:

$$\frac{Da_{ij}}{Dt} = -\frac{1}{2}[w_{ik}\,a_{kj} - a_{ik}\,w_{kj}] + \frac{1}{2}\lambda[\dot{\gamma}_{ik}\,a_{kj} + a_{ik}\,\dot{\gamma}_{kj} - 2a_{ijkl\dot{\gamma}_{kl}} + 2C_{l\dot{\gamma}}(\delta_{ij} - (2 + D_z a_{ij}))] \tag{2.3}$$

where D_z is the extra term called a thickness moment of interaction coefficient.

To improve the accuracy of fiber orientation predictions in simulations of injection-molded parts, the reduced strain closure (RSC) and anisotropic rotary diffusion (ARD) models have been implemented in a research version of Autodesk Moldflow Insight software. Overall, the RSC model captures the slow fiber orientation kinetics, which are observed in experiments but overpredicted by the widely used Folgar–Tucker model. While the ARD model accounts for fiber–fiber interactions using anisotropic diffusion to rectify the problem for long-fiber materials, the Folgar–Tucker model does not match all aspects of measured fiber orientation data [25].

2.6.2 Reduced strain closure model

The main objective of this model is to find the slow orientation dynamics. It is based on the concept of reducing the growth rates of the eigenvalues of the orientation tensor by a scalar factor, while leaving the rotation rates of the eigenvectors unchanged (eq. (2.4)) [24]:

$$\frac{Da_{ij}}{Dt} = -\frac{1}{2}\left[w_{ik}\,a_{kj} - a_{ik}\,w_{kj}\right] + \frac{1}{2}\lambda\left[\dot{\gamma}_{ik}\,a_{kj} - 2\left(a_{ijkl\dot{\gamma}_{kl}} + (1-\mathrm{k})\left(L_{ijkl} - M_{ijmn}\,a_{mnkl}\right)\right)\dot{\gamma}_{kl}\right.$$
$$\left. + 2kC_{l\dot{\gamma}}(\delta_{ij} - 3a_{ij})\right]$$

where

$$L_{ijkl} = \sum_{p=1}^{3} \sigma_p e_i^p e_j^p e_k^p e_l^p$$

$$M_{ijkl} = \sum_{p=1}^{3} \sigma_p e_i^p e_j^p e_k^p e_l^p$$

σ_p is the pth eigenvalue of the orientation tensor,
a_{ij} is the ith component and e_i^p the pth component; K is the phenomenological parameter. When $K \leq 1$, model the slow orientation dynamics and
$K = 1$, the RSC model is reduced to the original Folgar–Tucker model.

2.6.3 Anisotropic rotary diffusion model for long-fiber composites

The main objectives of this model are to calculate fiber orientation for long-fiber composite material. Long fibers are fibers longer than 1 mm, and to find the fiber–fiber interactions in long-fiber materials and predict all fiber orientation components simultaneously [24].

ARD replaces the isotropic diffusion via the following :

$$\frac{\mathrm{D}a_{ij}}{\mathrm{Dt}} = -\frac{1}{2}[w_{ik}a_{kj} - a_{ik}w_{kj}] + \frac{1}{2}\lambda[\dot{\gamma}_{ik}a_{kj} + a_{ik}\dot{\gamma}_{kj} - 2a_{ijkl}\dot{\gamma}_{kl} + \dot{\gamma}(2C_{ij} - 2C_{kk}a_{ij} - 5(C_{ik}a_{kj} + a_{kj}C_{ik}) + 10C_{kl}a_{ijkl}) \tag{2.5}$$

C_{ij} is assumed as a quadratic function of a_{ij} and $\dot{\gamma}_{ij}$ and is defined as

$$C_{ij} = b_1\delta_{ij} + b_2a_{ij} + b_3a_{ik}a_{kj} + b_4\frac{\dot{\gamma}_{ij}}{\dot{\gamma}} + b_5\frac{\dot{\gamma}_{ik}\dot{\gamma}_{kj}}{\dot{\gamma}^2}$$

where each is a scalar constant, and its values are determined by matching experimental steady-state orientation and requiring stable orientation.

Wang and Jin [24] compared the RSC (short-fiber material) and ARD (long-fiber material) models with the commonly used Folgar–Tucker model. They found that the two models reasonably predict the fiber orientation. On the other hand,Carton-Rose et al. [25] compared for the short glass fiber orientation the fiber orientation prediction provided by the RSC and Moldflow Adjusted Folgar–Tucker models within Autodesk Moldflow Insights software. Results show that for two-dimensional analyses for thin-walled moldings, the adjusted Folgar–Tucker model provides more suitable fiber orientation predictions when comparing with the RSC model. But for more complex components when three-dimensional analysis is required, the RSC model is more adequate.

2.7 Effect of fiber orientation on mechanical properties

According to the literature, several models have been developed to predict the material mechanical properties by taking into account several parameters such as fiber orientation, fiber length, fiber dispersion, fiber geometry and the degree of interfacial adhesion between fiber and matrix [26]. Indeed, all these parameters affect the material mechanical properties but it was reported that the "fiber orientation factor" and "fiber length factor" contribute the most to the strength of the composite [26].

The simpler are Voigt and Reuss models, which consist in a serial assembly (upper limit) or in a parallel assembly (lower limit) of the two phase materials. These models suppose that each phase has an isotropic elastic behavior.

Voigt assumed that the fibers and matrix are subjected to the same uniform strain in the fiber direction, while Reuss applied the same uniform stress on the fiber and matrix in the transverse direction (normal to the fiber direction). Both models can be described, respectively, by eqs. (2.6) and (2.7) [27, 28]:

$$E_V = V_f E_f + (1 - V_f)E_m \tag{2.6}$$

$$E_R = \frac{E_f\, E_m}{V_f E_m + (1 - V_f)E_f} \tag{2.7}$$

where E_V and E_R are, respectively, the Voigt and the Reuss elastic modulus, E_f and E_m represent the fiber and the matrix elastic modulus, respectively, and V_f is the fiber volume fraction.

Other models describe the composite as a combination of the Voigt and Reuss models. Hirsch model introduces into its equation (eq. (2.8)) an empirical parameter "x" ($0 < x < 1$) as a contribution of the two models ($x = 0.5$ random fiber orientation). It takes into account the effects of fiber orientation and stress concentration (fiber ends) [5]:

$$E_H = x\, E_V + (1 - x)E_R \tag{2.8}$$

where E_H, E_V and E_R are the Hirsch, Voigt and Reuss elastic moduli, respectively.

For its part, Tsai-Pagano model used values of 3/8 and 5/8 as a contribution of the longitudinal and transversal models, respectively (eq. (2.9)). It assumes a random filler orientation in the matrix as well as good dispersion and perfect fiber–matrix interfacial adhesion. It also includes the reinforcing fiber length and diameter [5]:

$$E_{T-P} = \frac{3}{8}E_V + \frac{5}{8}E_R \tag{2.9a}$$

$$E_{T-P} = \frac{3}{8} E_{H-T/L} + \frac{5}{8} E_{H-T/T} \tag{2.9b}$$

where $E_{T\text{-}P}$ is the Tsai-pagan elastic modulus.

On the other hand, Semlali et al. [5] developed a model based on a self-consistent approach using a modified version of the Hirsch model. Indeed this approach allows the material mechanical properties prediction by taking into account fiber orientation and content. The modification of Hirsch model consists of taking the empirical parameter (x) to be proportional to the fiber orientation factor (α), ranging between 0 and 1 to give a best fit as given in eq. (2.10). The self-consistent approach is based on the iterative insertion of a family of fiber orientation; hence, the first category of fibers was included in the matrix to constitute an equivalent homogeneous material, which in turn is integrated into an infinite composite:

$$E_h = \alpha\, E_V + (1 - \alpha) E_R \tag{2.10}$$

where E_h is the modified Hirsch elastic modulus.

Gupta et al. [7] used in their works the four models (Voigt, Reuss, Hirsch and Halpin-Tsai) to assume the tensile properties of a thermoplastic material-based wood fibers (L = 3.5 mm and d = 0.025 mm). To conclude: the parallel model gives the best prediction to the Young's modulus, while the Hirsch and Halpin-Tsai models predict a Young's modulus lower than the parallel model but higher than the experimental results. Conversely, the series model predicts the lowest value, which is also under the experiment results.

For its part, Semlali et al. [4] compared the predicted elastic modulus using the self-consistent approach, and Hirsch and Tsai–Pagano models of the PP/coir fiber biocomposites with the experimental one. Overall, their developed approach gives better results when comparing to Hirsch and Tsai–Pagano models.

2.8 Conclusion

The objective of this chapter was to contribute to the understanding and modeling of orientation phenomena in thermoplastic reinforced fibers during injection molding. In the first place, we have described the phenomenon of short fiber orientation during the flow of reinforced thermoplastics. This phenomenon is governed by mechanisms that induce a layer orientation of the fibers in the injected product, according to preferred directions, and this depends on the shape of the mold. Then, various digital software able to solve the problems related to the injection of composite were presented. It involves in particular succinctly describing the numerical methods used by REM 3D software and Moldflow insight for the resolution of the mechanical problem, typically here the resolution of the problem of the fiber orientation problem. Finally, since the mechanical properties of a composite are closely related to the fiber properties, namely orientation, several models of the composite mechanical performance prediction were presented.

References

[1] Mechraoui, A., Riedl, B., and Rodrigue, D. The effect of fibre and coupling agent content on the mechanical properties of hemp/polypropylene composites. Compos Interfaces 2007, 14, 837–848. doi: 10.1163/156855407782106591.

[2] Chang, R., and Yang, W. Numerical simulation of mold filling in injection molding using a three-dimensional finite volume approach. Int J Numer Methods Fluids 2001, 37, 125–148. doi: 10.1002/fld.166.

[3] Sengupta, R., Chakraborty, S., Bandyopadhyay, S., Dasgupta, S., Mukhopadhyay, R., Auddy, K. et al. A short review on rubber/clay nanocomposites with emphasis on mechanical properties. Polym Eng Sci 2010, 47, 2182–2189. doi: 10.1002/pen.

[4] Semlali Aouragh Hassani, F., Ouarhim, W., Zari, N., Bensalah, MO., Rodrigue, D., and Bouh, R. Injection molding of short coir fiber polypropylene biocomposites : Prediction of the mold filling phase. Polym Compos 2019, 1–14. doi: 10.1002/pc.25265.

[5] Semlali Aouragh Hassani, F., Ouarhim, W., Bensalah, MO., Essabir, H., Rodrigue, D., Bouhfid, R. et al. Mechanical properties prediction of polypropylene/short coir fibers composites using a self-consistent approach. Polym Compos 2018, 1, 1–11. doi: 10.1002/pc.24967.

[6] Bernasconi, A., Davoli, P., Basile, A., and Filippi, A. Effect of fibre orientation on the fatigue behaviour of a short glass fibre reinforced polyamide-6. Int J Fatigue 2007, 29, 199–208.

[7] Gupta, R., Sulaiman, N., and Gupta, A. An improved algorithm for prediction of Young's Modulus of wood plastic composites. Sci Res Essays 2013, 8, 649–656.

[8] Nguyen, QMP., Chen, X., Lam, YC., and Yue, CY. Effects of polymer melt compressibility on mold filling in micro-injection molding. J Micromech Microeng 2011, 21, 1–9.

[9] Fransisco, Folgar., and Charles, L. Tucker. III. Orientation behavior of fibers in concentrated suspensions. J Reinf Plast Compos 1984, 3, 98–119. doi: 10.1177/073168448400300201.

[10] Semlali Aouragh Hassani, F., Ouarhim, W., Zari, N., Bouhfid, R., and Qaiss, A el kacem. Natural fiber-based biocomposites 2019.

[11] Redjeb, A. Simulation numérique de l'orientation de fibres en injection de thermoplastique renforcé. Ecole Nationale supérieure des mines de Paris, 2007.

[12] Bay, RS., and Tucker, CL. Fiber orientation in simple injection moldings. part ii: Experimental results. Polym Compos 1992, 13, 317–331.

[13] Chung, ST., and Kwon, TH. Numerical simulation of fiber orientation in injection molding of short-fiber-reinforced thermoplastics. Polym Eng Sci 1995, 35, 604–618. doi: 10.1002/pen.760350707.

[14] COULON, A. Injection des polyamides renforcés de fibres de verre longues : Relations mise en oeuvre/comportement thermomécanique. 2008.

[15] Pontes, AJ., Neves, NM., and Pouzada, AS. The role of the interaction coefficient in the prediction of the fiber orientation in planar injection moldings. Polym Compos 2003, 24, 358–366. doi: 10.1002/pc.10035.

[16] Megally, A., and Etude, AM. Etude et modélisation de l'orientation de fibres dans des thermoplastiques renforcés. 2005.

[17] Shokri, P., and Bhatnagar, N. Effect of the post-filling stage on fiber orientation at the mid-plane in injection molding of reinforced thermoplastics. Phys Procedia 2012, 25, 79–85. doi: 10.1016/j.phpro.2012.03.053.
[18] Kenig, S. Fiber orientation development in molding of polymer composites. Polym Compos 1986, 7, 50–55. doi: 10.1002/pc.750070110.
[19] Gao, Y., Dong, X., Wang, L., Liu, G., Liu, X., Tuinea-Bobe, C. et al. Flow-induced crystallization of long chain aliphatic polyamides under a complex flow field : Inverted anisotropic structure and formation mechanism. Polymer (Guildf) 2015, 73, 91–101.
[20] Özdemir, A., Uluer, O., and Güldaş, A. Flow front advancement of molten thermoplastic materials during filling stage of a mold cavity. Polym Test 2004, 23, 957–966. doi: 10.1016/j.polymertesting.2004.04.011.
[21] Whiteside, BR., Coates, PD., Hine, PJ., and Duckett, RA. Glass fibre orientation within injection moulded automotive pedal – simulation and experimental studies. Plast Rubber Compos 2000, 29, 38–45. doi:10.1179/146580100101540680.
[22] Vincent, M., Giroud, T., Clarke, A., and Eberhardt, C. Description and modeling of fiber orientation in injection molding of fiber reinforced thermoplastics. Polymer (Guildf) 2005, 46, 6719–6725. doi: 10.1016/j.polymer.2005.05.026.
[23] Gupta, M., and Wang, KK. Fiber orientation and mechanical properties of short-fiber-reinforced injection-molded composites: Simulated and experimental results. Polym Compos 1993, 14, 367–382.
[24] Wang, J., and Jin, X. Comparison of recent fiber orientation models in Autodesk Moldflow Insight simulations with measured fiber orientation data. Proc Polym Process Soc 26th Annu Meet 2010.
[25] Caton-Rose, P., Hine, P., Costa, F., Jin, X., Wang, J., and Parveen, B. Fibre orientation and mechanical property predictions for short glass fibre reinforced injection moulding. Polym Process 2011, 1–5.
[26] Kalaprasad, G., Joseph, K., Thomas, S., and Pavithran, C. Theoretical modelling of tensile properties of short sisal fibre-reinforced low-density polyethylene composites. J Mater Sci 1997, 32, 4261–4267.
[27] Krishnan, KA., Anjana, R., and George, KE. Effect of alkali-resistant glass fiber on polypropylene/polystyrene blends : Modeling and characterization. Polym Compos 2014, 1–9. doi: 10.1002/pc.
[28] Hu, H., Onyebueke, L., and Abatan, A. Characterizing and modeling mechanical properties of nanocomposites- review and evaluation. J Miner Mater Charact Eng 2010, 9, 275–319. doi: 10.4236/jmmce.2010.94022.

Section II: **Characterization and Properties**

Raghu Raja Pandiyan Kuppusamy

3 Characterization of an unsaturated polyester resin for liquid composite molding processes

Abstract: Liquid composite molding (LCM) process involves different phenomena such as resin flow, heat transfer and polymerization reactions simultaneously. Mold filling and subsequent curing are the significant processing stages that need to be modeled in LCM process simulation. Different submodels quantifying flow, heat and mass transfer are required for the main model to describe LCM process filling and subsequent curing stages realistically. The resin polymerization reaction leads to the phase transformation from viscous liquid to rigid solid with an exothermal effect. The gel time, which marks the onset of viscous resin liquid to gel stage, constitutes a crucial parameter for the mold fill time. The required component geometry needs to be filled before the resin gels. The nature of flow progression is affected by heat and cure reaction through the resin viscosity, because the resin viscosity changes with the temperature and the extent of cure reaction. Eventually, the rate of cure reaction, degree of cure and the nature of exothermic heat evolved depend on the applied temperature. Hence, the models depicting resin cure reaction kinetics and viscosity as a function of temperature and degree of cure constitute the submodels to the main flow, heat and mass transfer models. In this work, general-purpose unsaturated polyester resin was characterized for gelation and exotherm behavior, resin cure kinetics, resin cure viscosity for applications in LCM process simulations.

Keywords: Liquid composite molding, resin gelation, cure kinetics, cure viscosity, resin exotherm

3.1 Introduction

Liquid composite molding (LCM) processes such as resin transfer molding (RTM) are versatile and effective means for producing complex and large composite structures. Development and optimization of RTM process requires a proper mold design, a successful injection strategy and a well-defined resin cure cycle. Experimental studies based on experience and trial-and-error methods are often expensive and time-consuming. Alternatively, computer simulation of the molding process can predict

Raghu Raja Pandiyan Kuppusamy, Department of Chemical Engineering, National Institute of Technology, Warangal, India

https://doi.org/10.1515/9783110655049-003

flow fronts, allow engineers to virtually test different mold designs without building expensive hardware. RTM process consists of three phases: fiber preforming, mold filling and the resin cure cycle. The mold filling and the cure phase are considered independent, since the characteristics of the RTM flow process (resin viscosity, applied pressure or flow rate at the injection gates, gate and vent locations, etc.) are usually preferred in such a way that the mold filling phase is completed before the resin gelation and the curing phase starts after the mold filling phase is completed. In other words, resin properties such as resin gel time, resin viscosity and resin cure behavior are the prerequisites for the prosperous RTM mold fill and cure phases.

A number of studies have been posted on gelation and exotherm behavior of unsaturated polyester (UP) resins. There have been a few, more detailed, analyses of the effect of the concentration of initiator and accelerator on the cure behavior of polyester resins [1]. A range of viscosity empirical models has been reported for the viscosity evolution up to resin gelation. For isothermal situations, an alternative method has been proposed in modeling the viscosity advancement as a function of elapsed cure time [2]. Many researchers have studied the cure kinetics of UP resin by different techniques and have found that differential scanning calorimetry (DSC) by isothermal mode is the predominant technique to read the resin cure kinetics [3]. Despite the extensive literature on the curing of thermoset resins, detailed studies on specific resin systems particularly designed for specific applications are generally not available

In general, UP resins can be formulated as high, medium and slow reactive depending on the density of reacting sites that are tailored to suit different applications such as hand lay-up, LCM processes including RTM and pultrusion process. However, the choice of particular reactivity UP resin for a particular manufacturing process depends on a balance between wetting time and cure time. The slow reactivity UP resin is preferred for hand lay-up process as its gel time typically ranges from several minutes to hours, whereas medium reactivity UP resin is preferred for LCM processes including RTM since its gel time falls in the order of minutes.

In this work, the UP resin was characterized for the following parameters to aid the RTM process simulations:
- Gel time and exotherm behavior using ASTM method
- Resin cure kinetics using DSC method
- Viscosity evolution during curing using viscometer

The materials used, methods adopted and the results obtained for the characterization of UP resin are explained below.

The main objective of this work is to find out physical and design parameters of an UP resin using appropriate characterization techniques so as to portray the real RTM process simulation. This chapter is structured as follows: First, the effect of concentration of initiator and accelerator has been examined on gel time variation and cure behavior for UP resin. Then, the resin cure viscosity profiling with cure time dependency has been presented. Finally, extension of cure chemical reaction is quantified using classic resin cure kinetic model-based fitting technique.

3.2 Materials and experimental methods

The study was based on isophthalic UP resin (Grade: Mechester 1110M medium reactivity resin, Make: Mechemco Industries, India) as polymer matrix. Methyl ethyl ketone peroxide and cobalt naphthanate were used as initiator and promoter, respectively. UP resin, being the prime raw material, has been characterized for its gel time, viscosity and cure behavior. The gel time, peak exothermic temperature and peak exotherm were determined by the following ASTM D2471 [4]. The gel time test and curing studies were performed for different combination catalyst and accelerator volumes. In this chapter, accelerator volume varied from 0.5 to 2.5 mL and the catalyst volume varied from 0.5 to 2.5 mL per 100 mL of resin.

Initially, 100 mL of the resin sample was added with selected proportion of accelerator and it was agitated slowly with a stirrer avoiding air entrapment. Start the timer when the known volume of catalyst was added. To avoid transfer of heat, do not hold the container by hand during the mixing operation. A temperature measuring device was inserted into the geometric center of the reacting mass, and observed temperature changes were recorded to the end of the test. Every 15 s, the center surface of the reacting mass was probed with the applicator stick perpendicular to the material surface. When the reacting material no longer adheres to the end of a clean probe, "gel time" was recorded as the elapsed time from the start of mixing of catalyst. The time and temperature were recorded continuously until the temperature starts to drop. Highest temperature reached was recorded as the "peak exothermic temperature" and the elapsed time from the start of mixing to reach the maximum temperature is recorded as "peak exothermic time."

The flow behavior of the resin is very important for controlling the production process via the resin injection and fiber wetting. Viscosity is the chief property which governs the flow of the resin. The resin sample was tested for viscosity using Bohlin viscometer. The resin was tested with single shear as well as with variation of shear. The goal of single shear was to find the change of viscosity with time, whereas variation of shear was done to find the change in viscosity with change in shear. Viscosity of the catalyzed resin before the gel time was also measured. The reacting resin sample was tested for viscosity using Bohlin viscometer [5]. The resin sample was introduced with single shear stress and single shear rate of magnitude 78 Pa and 60 per secs every 5 min at the environment of 25.8 °C temperature.

The DSC analysis was carried out for cure kinetics analysis with a small mass (10–15mg) of sample using a Perkin-Elmer DSC-7 instrument operating with nitrogen atmosphere. All DSC measurements were done in hermetic aluminum pans.

The curing of these samples was performed under isothermal condition. To select suitable temperatures for the required isothermal experiments, a dynamic DSC run at a heating rate of 10 °C/min was first performed. Temperatures above but near the onset of reaction and below the peak heat flow were chosen for the isothermal cure study. Temperatures above but near the onset of reaction were chosen. In this work, isothermal experiments were carried out at 100, 110 and 120°C.

3.3 Results and discussions

3.3.1 Resin gelation and exotherm behavior

The results obtained from resin curing experiment as per ASTM D2471 standard are given in Table 3.1. The effect of catalyst and accelerator concentration on resin gelation and exotherm behavior are tabulated in Table 3.1.

Table 3.1: Effect of accelerator and catalyst concentration on resin cure exotherm.

S. no.	Resin volume (mL)	Accelerator volume (mL)	Catalyst volume (mL)	Gel time (s)	Peak exothermic temperature (°C)	Peak exotherm temperature time (s)
1	100	2.5	2.5	227	167	750
2	100	2.0	2.5	238	167	765
3	100	1.5	2.5	249	168	825
4	100	1.0	2.5	384	169	1,055
5	100	0.5	2.5	1,151	169	2,055
6	100	2.5	2.0	244	159	780
7	100	2.0	2.0	290	160	840
8	100	1.5	2.0	360	160	915
9	100	1.0	2.0	416	160	1,020
10	100	0.5	2.0	1,213	160	2,220
11	100	2.5	1.5	248	157	750
12	100	2.0	1.5	305	156	765
13	100	1.5	1.5	372	156	825
14	100	1.0	1.5	448	156	1,055
15	100	0.5	1.5	2,705	155	2,055
16	100	2.5	1.0	754	152	1,410
17	100	2.0	1.0	975	152	1,665
18	100	1.5	1.0	1,210	151	2,055
19	100	1.0	1.0	1,655	152	2,730
20	100	0.5	1.0	6,422	150	9,765

Table 3.1 (continued)

S. no.	Resin volume (mL)	Accelerator volume (mL)	Catalyst volume (mL)	Gel time (s)	Peak exothermic temperature (°C)	Peak exotherm temperature time (s)
21	100	2.5	0.5	1,564	147	2,550
22	100	2.0	0.5	1,776	147	2,865
23	100	1.5	0.5	2,467	146	4,170
24	100	1.0	0.5	5,028	145	8,265

Gel time will be the check parameter for mold filling simulation, which determines the catalyst and accelerator proportions. From Table 3.1 and Figure 3.1, it is evident that, for a particular catalyst proportion, the gel time is reduced as the amount of accelerator increases. On the other hand, for a particular accelerator proportion, the gel time shows the same trend with increase in catalyst. This supports the selection of relevant proportion of accelerator and accelerator for a particular mold fill time.

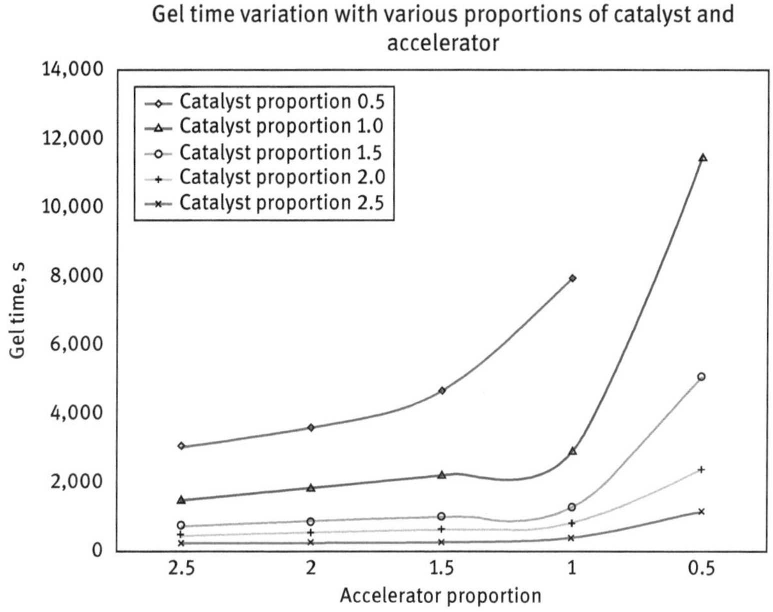

Figure 3.1: Gel time variation with various catalyst and accelerator proportions.

The time–temperature curve at constant applied cure temperature for aforementioned combinations of catalyst and accelerator was shown to acquire the nature of rise in temperature and to find out minimum time required for the cure for a specific combination catalyst and accelerator proportions.

Figure 3.2 shows that cure cycle time was less for higher proportions of catalyst and accelerator which is the most considerable factor for production time, whereas the peak exothermic temperature was found more with increasing catalyst proportion. Similar graphs can be produced for catalyst proportion variation from 2 to 0.5. The peak exothermic temperature varies from 169 to 145 °C with the catalyst volume differs from 2.5 to 0.5 mL per 100 mL of resin, as shown in Figure 3.3. It is more advantageous to have less peak exothermic temperature, since higher peak exothermic temperature can cause shrinkage, warping, residual stress, degradation of polymer, smoke, resin cracking and so on.

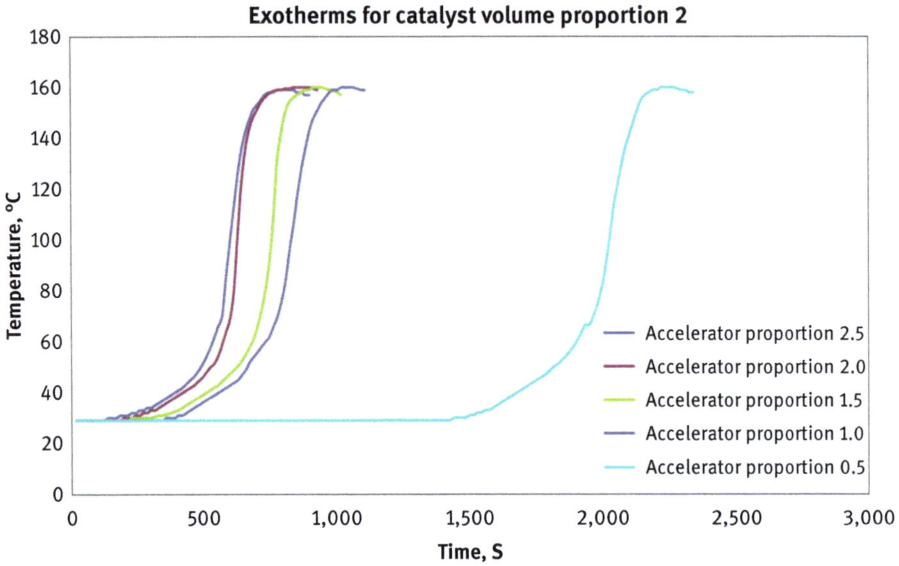

Figure 3.2: Exotherms for catalyst proportion 2.5 with various accelerator proportions.

3.3.2 Resin cure kinetics

The task of the cure kinetics modeling is to uniquely define the curing reaction rate with its variables. The rate of reaction $d\alpha/dt$ is measured by the product of two functions, namely, temperature-dependent function $k(T)$ and cure conversion-dependent function $f(\alpha)$. The temperature-dependent function $k(T)$ is given by the Arrhenius form. The dependence of the reaction rate on the extension of reaction takes the form as follows:

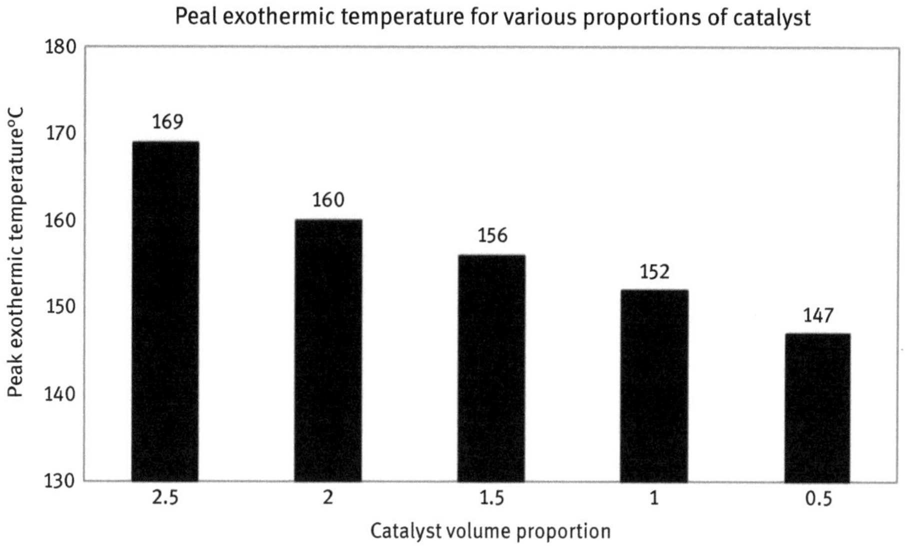

Figure 3.3: Peak exothermic temperature for varying catalyst proportion.

$$f(\alpha) = (1 - \alpha)^n \tag{3.1}$$

DSC measures the amount of heat generated during curing reaction. The heat flow versus time was measured at the isothermal temperatures until the reaction was complete. The extent of cure reaction (or) the degree of cure at time t $\alpha(t)$ can be determined from the measured DSC heat flow curves using eq. (3.2):

$$\alpha(t) = \frac{\Delta H_t}{\Delta H_{\text{tot}}} \tag{3.2}$$

For a given isothermal temperature, ΔH_t is the accumulative heat of reaction up to curing time t and given by area under the heat flow curve at time t. ΔH_{tot} is the total enthalpy of the reaction and given by the total area under the heat flow curve. The steps involved in the modeling of cure kinetics are as follows [6]:

i. Obtain heat flow versus time curve using DSC experiment for the applied isothermal temperatures.
ii. Numerically integrated heat flow versus time curve to obtain the total heat of reaction ΔH_{tot}.
iii. Obtain partial reaction heat ΔH_tat time t by partial numerical integration of heat flow versus time curve.
iv. Obtain degree of cure $\alpha(t)$, using eq. (3.2) and produce degree of cure versus time curve.
v. Obtain the rate of degree of cure $d\alpha/dt$ by numerically differentiating the degree of cure versus time curve and produce rate of degree of cure versus time curve.

vi. Obtain the rate of degree of cure as a function of degree of cure for the applied isothermal temperatures with time as a parameter.
vii. Using nonlinear regression analysis, fit rate of degree of cure as a function of degree of cure with the classic resin cure kinetics model
viii. Obtain cure rate constant k for the applied isothermal temperatures and fit the temperature dependency of rate constant using Arrhenius form.

In this work, modified Kamal and Sourour cure kinetics model has been utilized to quantify the rate of degree of cure as a function of degree of cure and temperature. The comparisons for isothermal 100, 110 and 120 °C experimental cure rate versus degree of cure data were given in Figures 3.4–3.6. It may be mentioned that the model parameters in the modified Kamal and Sourour kinetics model have been evaluated by using nonlinear regression analysis to our experimental data; the values of the parameters for the modified Kamal and Sourour kinetics model as well as the correlation factors for the fits are presented in Figures 3.4–3.6. The relationship between cure rate constant determined from Kamal model fit and cure temperature was linearized using functional form of Arrhenius equation, as shown in Figure 3.7. The determined frequency factor and activation energy are 3.71×10^6 min^{-1} and 64.43 kJ/mol.

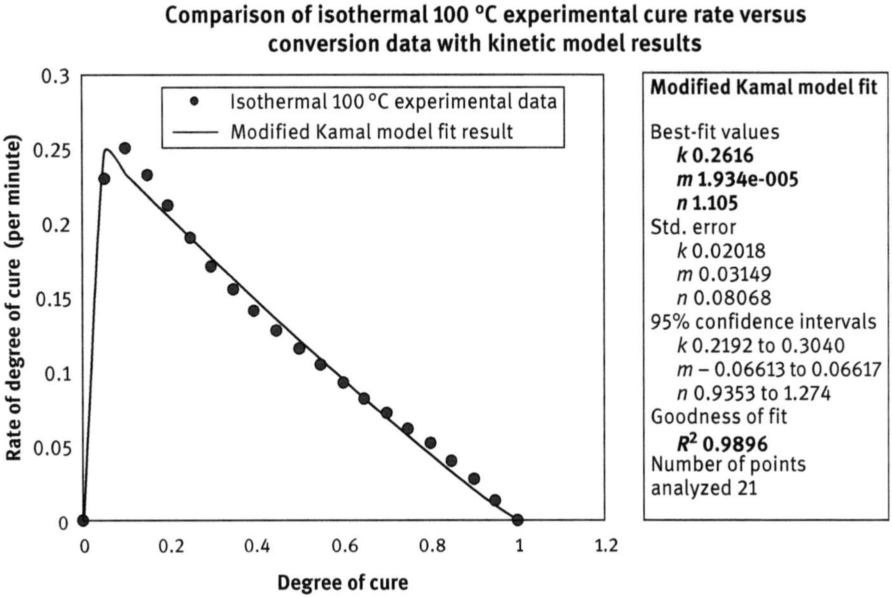

Figure 3.4: Comparison of experimentally obtained rate of degree of cure versus degree of cure at 100 °C with that predicted by the modified Kamal and Sourour kinetics model.

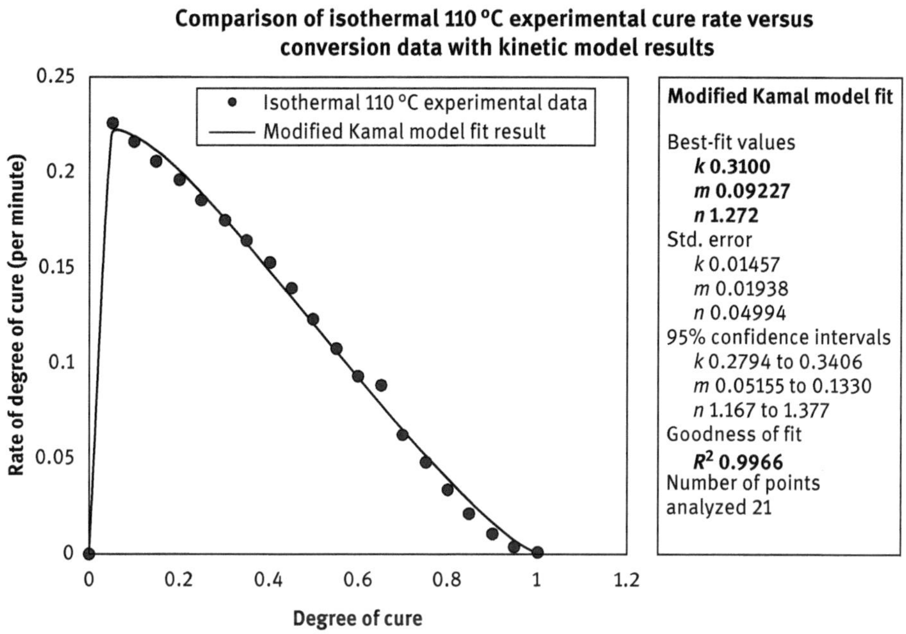

Figure 3.5: Comparison of experimentally obtained rate of degree of cure versus degree of cure at 110 °C with that predicted by the modified Kamal and Sourour kinetics model.

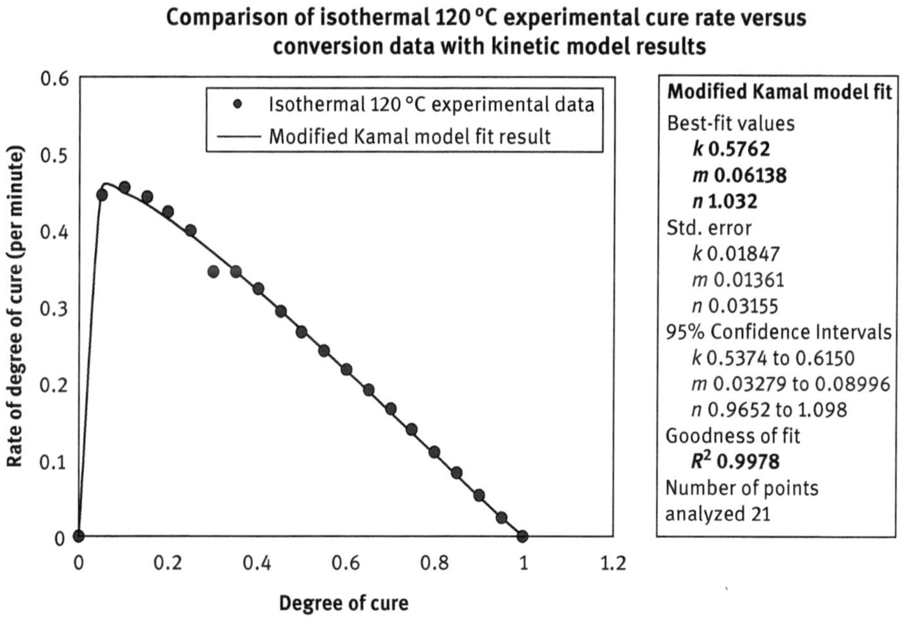

Figure 3.6: Comparison of experimentally obtained rate of degree of cure versus degree of cure at 120 °C with that predicted by the modified Kamal and Sourour kinetics model.

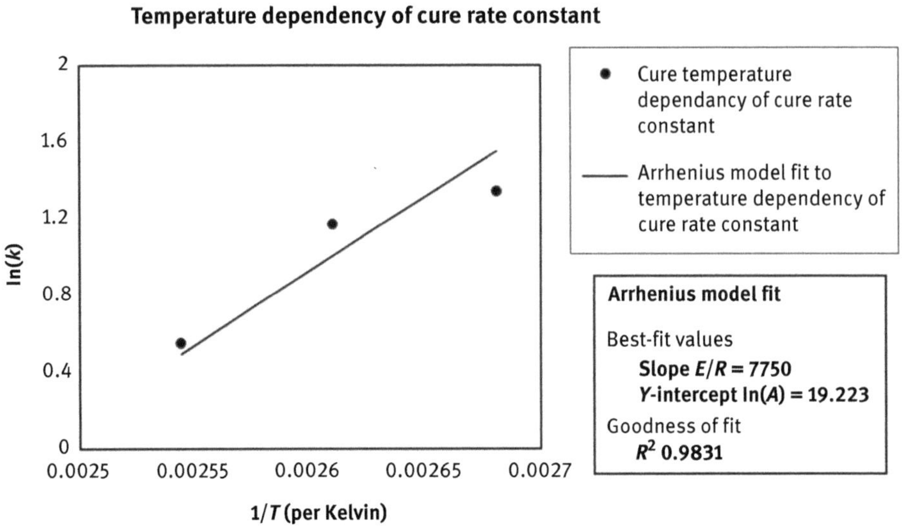

Figure 3.7: Temperature dependence of the cure kinetic rate constants.

3.3.3 Resin cure viscosity

The raw resin viscosity measurements using single shear and table of shear are given in Tables 3.2 and 3.3. From the single shear table, it is evident that there is no appreciable change in viscosity due to time factor and it shows a steady-state viscosity of 1.30 Pa s. In case of table of shears, the resin shows Newtonian nature with the magnitude of 1.26 Pa s.

Table 3.2: Raw resin viscosity using single shear analysis.

Time '(s)	Temperature '(°C)	Shear rate '(1/s)	Shear stress '(Pa)	Viscosity '(Pa s)
10	25.8	60.74	78.81	1.2975
25.062	25.8	60.73	78.73	1.2963
40.343	25.8	60.73	79.32	1.3060
55.359	25.8	60.73	79.11	1.3027
70.421	25.8	60.73	78.82	1.2979
85.484	25.8	60.72	78.88	1.2991
100.546	25.8	60.72	79.23	1.3048
115.609	25.7	60.72	79.15	1.3036
130.671	25.7	60.72	79.51	1.3095
145.734	25.7	60.72	79.51	1.3095

Table 3.3: Raw resin viscosity using table of shear analysis.

Time '(s)	Temperature '(°C)	Shear rate '(1/s)	Shear stress '(Pa)	Viscosity '(Pa s)
15.047	25.7	19.46	25.09	1.28930
27.313	25.7	42.31	56.08	1.32520
38.828	25.7	65.38	85.76	1.31160
49.953	25.7	88.33	114.34	1.29440
60.907	25.7	111.67	142.92	1.27980
71.844	25.6	134.55	170.99	1.27080
82.625	25.6	157.83	200.29	1.26910
93.219	25.6	180.96	228.81	1.26440
103.782	25.6	204.42	257.94	1.26180
114.282	25.6	227.70	287.18	1.26120
124.719	25.6	251.02	315.15	1.25550
135.157	25.6	274.41	343.32	1.25110
145.594	25.5	298.14	371.51	1.24610
155.969	25.5	321.15	399.35	1.24350
166.344	25.5	344.61	427.12	1.23950
176.719	25.5	363.37	449.35	1.23660
187.032	25.5	386.72	476.85	1.23310
197.344	25.5	410.49	504.20	1.22830
207.657	25.5	433.78	530.65	1.22330
217.985	25.4	457.42	557.65	1.21910

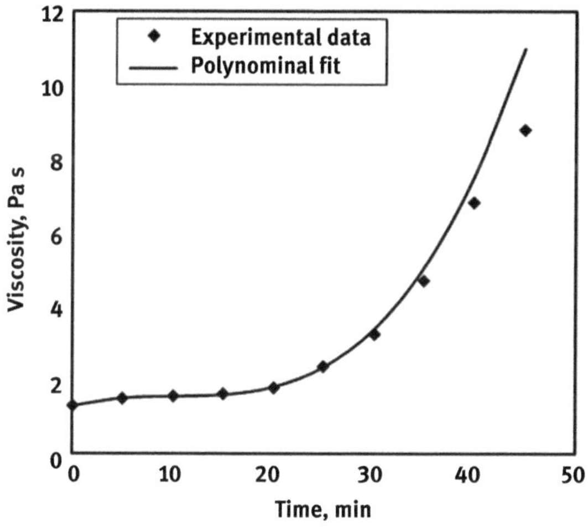

Third-order polnomial fit
Best-fit values
B0 1.304
B1 0.06866
B2 −0.006488
B3 0.0002166
Std. error
B0 0.03751
B1 0.01457
B2 0.001448
B3 3.804e-005
95% confidence intervals
B0 1.143 to 1.465
B1 0.005967 to 0.1314
B2 −0.01272 to −0.0002564
B3 5.293e-005 to 0.0003803
Goodness of Fit
R^2 0.9956
Number of points analyzed 10

Figure 3.8: Isothermal resin cure viscosity.

The characteristic viscosity profile and its mathematical modeling given in Figure 3.8 are obtained from an isothermal cure at room temperature. During an isothermal cure, the viscosity of the resin increases continuously as a result of cross-linking. Thus, in a viscosity versus cure time plot, the sudden increase in viscosity will mark the onset of gelation.

3.4 Conclusions

The characterization of specific grade of an UP resin has been performed for its gel time, cure behavior, cure kinetics and cure viscosity. The results obtained have thus been utilized to aid RTM mold fill and cure phases.

References

[1] Hossein Beheshty, M., Nasiri, Hassan., and Vafayan, Mehdy. Gel time and exotherm behaviour studies of an unsaturated polyester resin. Iranian Polymer Journal 2005, 14(11), 990–999.

[2] Karkanas, Panagiotis I., and Partridge, Ivana K. Cure modeling and monitoring of epoxy/amine resin systems. II. Network formation and chemoviscosity modeling. J Appl Polym Sci 2000, 77, 2178–2188.

[3] Du, Shanyi., Guo, Zhan-Sheng., Zhang, Boming., and Wu, Zhanjun. Cure Kinetics of epoxy resin used for advanced composites. Polym Int 2004, 53, 1343–1347.

[4] Kuppusamy, Raghu Raja Pandiyan, and Neogi, Swati. Influence of curing agents on gelation and exotherm behaviour of an unsaturated polyester resin. Bull Mater Sci 2013, 36(7), 1217–1224.

[5] Kuppusamy, Raghu Raja Pandiyan, and Neogi, Swati. Viscosity modeling of a medium reactive unsaturated polyester resin used for liquid composite molding process. J Appl Polym Sci 2012, 125, 1400–1408.

[6] Kuppusamy, Raghu Raja Pandiyan, Chakraborty, Saikat., Kundu, Gautam., and Neogi, Swati. Curing kinetics of medium reactive unsaturated polyester resin used for liquid composite molding process. J Appl Polym Sci 2009, 114, 2415–2420.

Emmanuel Baffour-Awuah, Stephen Akinlabi and Tien-Chien Jen

4 Properties of oil palm lignocellulose fiber and polymer composite: two-and-a-half decade overview

Abstract: The aim of the chapter was to review the characteristics of oil palm fiber (OPF)–polymer composites or blends. Nonetheless, the objectives were to dilate on the effects of OPF involving empty bunch fiber and oil palm mesocarp fiber on selected polymer materials. Polymers considered include natural rubber, polyethylene, polypropylene, polyvinyl chloride, polystyrene, polyurethane, polyester and epoxy. Documents mainly relating physical, mechanical, thermal and electrical characteristics of OPF–polymer blends were considered. The chapter portrays mixed effects. Thus while some polymer (matrix) properties enhance as a result of filler (fiber) presence, others retrogress. As a composite material, OPF–polymer materials, therefore, exhibit different characteristics based on five factors. These are fiber type (bunch or mesocarp); fiber size, fiber percentage weight or volume in composite; fiber percentage volume in composite; and matrix type. Therefore, there is a great potential for studies in this area of composite development. With respect to characterization, the chapter provides an assemblage of knowledge and information for prospective researchers as a one-stop medium toward studies relating OPF–polymer composite formation. Composite OPF–polymer hybridization is a technique through which overall composite characteristics may be enhanced. Reviewed literature suggests that future researchers may find this area of studies exciting and innovative since it appears very scanty work, if any, concerning OPF–polymer hybridization has been pursued in this regard.

Keywords: Bioplastic, composite formation, hybrid composites, OPF, polymer

Emmanuel Baffour-Awuah, Department of Mechanical Engineering Science, Faculty of Engineering and the Built Environment, University of Johannesburg, Auckland Park, South Africa. Department of Mechanical Engineering, School of Engineering, Cape Coast Technical University, Cape Coast, Ghana.
Stephen Akinlabi, Department of Mechanical & Industrial Engineering Technology, Faculty of Engineering and the Built Environment, University of Johannesburg, Auckland Park, South Africa. Department of Mechanical Engineering, Covenant University, Ota, Nigeria
Tien-Chien Jen, Department of Mechanical Engineering Science, Faculty of Engineering and the Built Environment, University of Johannesburg, Auckland Park, South Africa.

https://doi.org/10.1515/9783110655049-004

4.1 Introduction

For more than four decades, global attention on environmental sustainability has attracted application of natural fibers for various purposes. Among those applications is utilization of biodegradable materials in the formation of composite materials [1]. In civil and construction, consumer packaging, aircraft and automotive industries among others, natural fiber–polymer composites are gradually becoming alternative materials as a result of a number of benefits and advantages, particularly to the environment when compared with traditional nonbiodegradable composites.

Though plastics may be manufactured from nonbiodegradable materials, the inclusion of natural fiber in composite formation goes a long way to reduce the environmental burden with reference to homogeneity of pure nonbiodegradable polymer products [2, 3]. For this reason, natural fibers such as jute, kenaf, sisal, cotton, flax, banana, pineapple leaf and oil palm fiber (OPF) are very much in focus in this regard as far as biocomposite manufacture is concerned [4, 5]. Among the variety of natural fibers as a choice of biopolymer composites, OPF appears to be a preferable choice in many tropical regions of the globe in general, and particularly in areas where climate supports cultivation of the plant. There is no doubt, therefore, that within nations located within tropical regions, research is still ongoing in ascertaining the potential of OPF as a waste material in the development of applicable engineering materials that are capable of sustaining continuous reliability of the global environment.

Fundamentally, two technical techniques are employed in the development of biodegradable plastic composites based on the type of parent material. Parent material may be biodegradable or nonbiodegradable. Nonbiodegradable plastics are manufactured from petroleum while biodegradable plastics are produced from plants. Plastics may be thermoset-based or thermoplastic-based. Examples of thermosets include phenol formaldehyde resin, epoxy (EP) and unsaturated polymers. Thermoplastic polymers include polyvinyl chloride (PVC), polyethylene (PE), polystyrene (PS), polyurethane (PU) and polypropylene (PP). The obdurate nature of thermosets, which involve high processing temperature, lengthy cure cycles, brittleness, irreparability and nonrecyclability, has rather made thermoplastics preferable to polymer–natural fiber composite industry and research [6]. Biodegradable plastics produced from plants include starch, cellulosic plastics, soybean plastics, polycarpic acids, polyhydroxybutyrate, polycaprolactone and polyhydroxyalkanoate, which may also be formed as composites with natural fiber to form natural fiber–polymer composites [7–9].

A composite material may be described as a dual-phase material made of a continuous phase and a dispersed phase; the continuous phase occupying the interfibrous volume and thus acting as a load transfer medium to the dispersed phase, while the dispersed phase also tends to build upon the property(ies) of the entire composites [10]. Biocomposites are formed when a biodegradable or petroleum-based nonbiodegradable plastic combines with a biobased fibril. Biodegradable polymers are derived from crop/bioderived plastic also referred to as bioplastic or biopolymer. Since they are friendlier to ecological systems and the environment, they are usually referred to as green composites [11].

Manufacturing of composites begins with the introduction of synthetic fibers as reinforcement materials. Carbon glass and aramid fibers were the earlier synthetic fibers, which were employed since their composites yield specific variety of engineering properties. Over the past three decades, use of natural fiber as reinforcement materials has also become popular in both research and manufacturing industries. Natural fiber composites have manifold advantages. They have higher deformability and biodegradability [12]; contribute positively to global carbon budget [13]; have high ease of recyclability and combustibility [14]; are low in cost; are less abrasive to manufacturing processes; have low density as well as ecologically and environmentally friendly [15]. In addition to these advantages, natural fiber composites have good thermal and insulation properties [16], high specific strength, flexibility in applications, better nontoxicity and better renewable nature [17]. Natural fiber can also be composted for agricultural applications; the life cycle benefits which include calorific utilization is far greater than the synthetic polymer composites. They have good acoustic insulation and electrical resistance properties, and are widely dispersed over the globe, thus easily available and accessible [18]. Natural fiber has further been found to have high mechanical performance and good dimensional stability.

In spite of the merits associated with natural fibers, the performance of their composites, to a large extent, depends on the fiber–matrix bond and the suitability of selected process variables [19–34]. When a fiber–polymer interface is well bonded, there is adequate stress dispersion between the plastic and fibril, thus ensuring a more quality fiber-reinforced composite [25, 26, 29]. Natural fibers are, however, hydrophilic naturally. In order to circumvent this deficiency as a reinforcement material, they are modified through various techniques. These include plasma, physical (thermal treatment, stretching or calendering), biological and chemical applications [24, 25, 35].

Interest on natural fiber composites in various engineering applications and research has been growing because of the numerous advantages. Sreekala et al. (2002) reported fiber-reinforced plastic composite applications in industrial and electronic machinery, sports and recreation equipment as well as aerospace technology. Other applications include marine technology; automotive components as well as office and stationery and educational technology. John and Thomas [11] also attest to natural fiber composite application by Daimler Chrysler and Mercedes Benz, whereby flax-sisal fiber mat matrixed with EP is used to manufacture door panels of Mercedes Benz E-class model. Seatings of Mercedes Benz A-class brand are also produced from bonded coconut fiber and natural rubber (NR) latex. Rear shelf trim panels of 2000 Chevrolet Impala model and some automotive components of some industries in the United States, for example the Cambridge Industry, are employing flax fiber–PP composite materials. In the building and construction industry, lignocellulosic fiber composites have found applications in production of partition boards, panels and ceiling as well as block and floor work [16, 36].

Natural fibers may be categorized in terms of source or origin. Thus they may be originated from plants, animals or minerals, as against synthetic fibers that are man-made in nature. Synthetic fibers are produced by spinning synthetic polymers (chemically produced). Most natural fibers employed as reinforcing materials are plant based. Lignocellulosic or plant fibers include wood (softwood or hardwood), stem or bast (flax, hemp, ramie, jute, kenaf), leaf (sisal, pineapple, fique, palm, abaca, banana, henequen), seed or fruit (cotton, coconut, oil palm) and grass (bamboo, rice) [37]. Figure 4.1 shows the nature of OPFs as part of the family tree of natural fibers.

Oil palm is one of the most popular natural fiber plants in the world for two reasons: first, it is the highest edible oil yielding plant; second, wood, leaf, fruit and seed are all fibrous in nature [37]. World production of palm oil hit 21.8 million tons in 2000 with Malaysia producing about half of the produce (Basiron, 2002). In 2013, the total land coverage of oil palm plantation was about 5.23 million hectares in Malaysia, with oil palm and palm products contributing 8.8% of export earnings [38, 39]. Globally, 11 million hectares of land was cultivated in about 42 countries [40].

Countries that are prominently featured in oil palm production in West Africa include Ghana, Cote Divoire and Nigeria; Malaysia and Indonesia in Southeast Asia; India in Southern Asia; as well as Brazil in South America [41, 42]. While up to an annual production of palm oil is obtained as 5.5 tons, fibrous biomass of dry matter may amount to 55 tons from a hectare of oil palm plantation [43]. Thus, a substantial quantity of fibrous biomass is lost as waste in farms and milling sites. As already indicated, lignocellulosic fibers can be obtained from trunk, frond, empty fruit bunch (EFB) and fruit mesocarp of the oil palm tree. Fibrous biomass is disposed within plantation farms and manufacturing plant environments close by. Tons of fibrous materials are thus produced annually as waste and to the detriment of the environment [44]. However, other applications have been innovated. These include application as boiler fuel [45] and mulching material or fertilizer application [46]. In spite of these applications, tons of OPF is still left in plantations to rot, eventually causing environmental nuisance and pollution. In view of its abundance and possible usage, both industry and researchers have developed intense interest in research work involving bioplastic composite applications of the fiber with particular reference to utilization and characterization within the last five decades or so [47].

The aim of the chapter was to identify the effects of OPF size and weight on properties of OPF–plastic blends produced through bonding of natural fiber with thermosetting polymers. However, objectives were (1) to identify the effect of OPF on OPF–plastic blends; and (2) to identify the effect of parent thermoplastic matrix type on OPF–polymer blends. For the purpose of this chapter, the review was limited to OPFB (oil palm fruit bunch) fiber and oil palm mesocarp fiber (OPMF) as vegetable

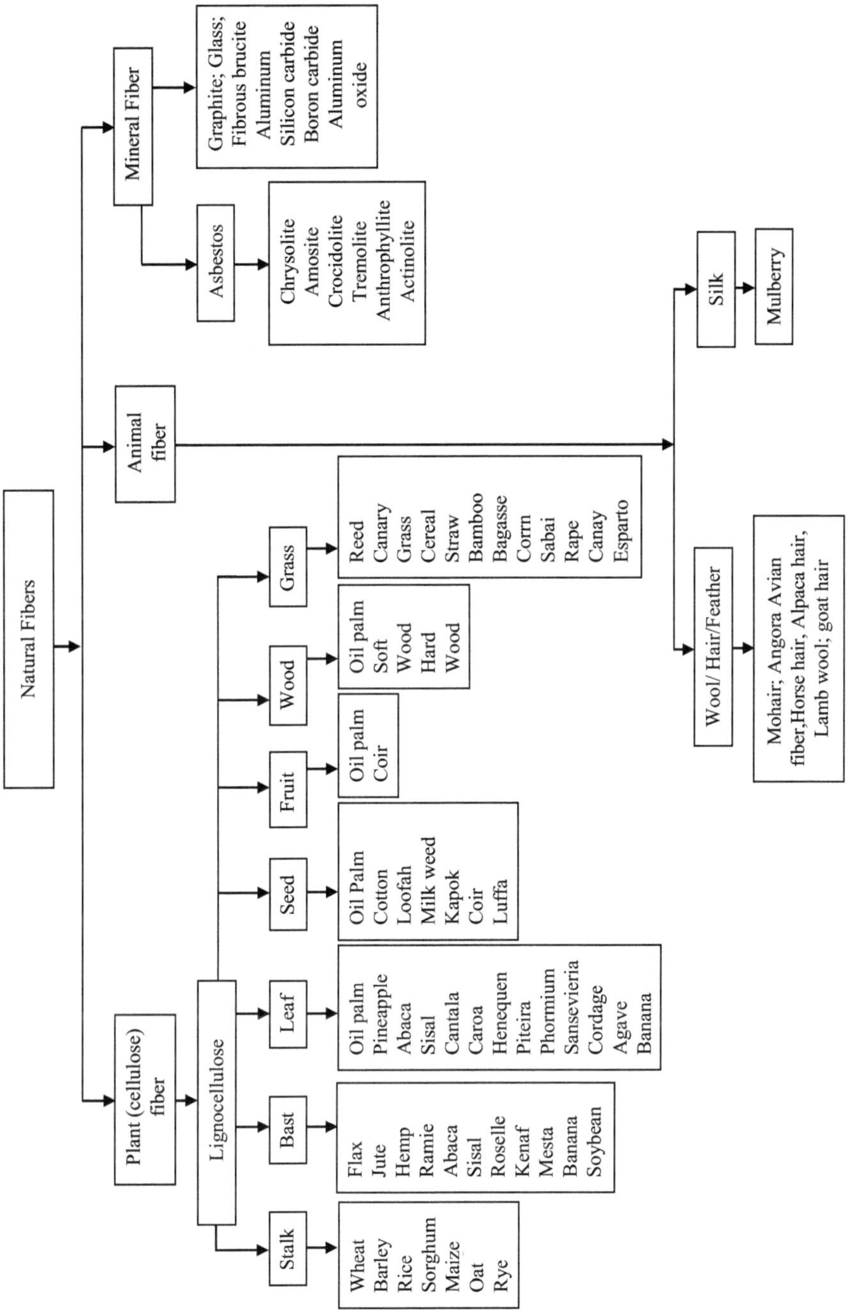

Figure 4.1: The ancestral tree of oil palm in the family of natural fiber.

fibers. Thermoplastic polymers that were considered include NR, PP and PU. The remaining were PVC, polyester (PES) and PS.

4.2 Oil palm fiber

The OPF is an ancillary product obtained during processing of palm oil. Palm oil is produced from the palm fruit, which in turn is obtained from the palm tree. Palm trees are cultivated either on farms and small-, medium- and large-scale plantations or found in the wild. Quality of palm oil in terms of field conditions largely depends on the age of palm tree, environment conditions, genetic influence, agronomic practices, harvesting methods and handling and transport factors. The fresh fruit may be obtained either as loose and single or as a bunch. If received as a bunch, fruits embedded in spikelets which grow on a trunk are removed by threshing.

Threshing is a manual process whereby fruit-embedded spikelets are cut from bunch stem with a machete or an axe. Fresh fruits may also be removed mechanically using a rotating drum and rotary beater mechanism, the spikelets being left on the bunch stem. Threshing may be introduced either after cooking or sterilizing the bunch. This facilitates easy detachment of the fruits from the spikelets. Sterilization or cooking is the application of high-temperature wet heat treatment to loosen the fresh fruits. While sterilization employs pressurized steam, cooking uses hot water. The fiber in the EFB is referred to as the oil palm empty fruit bunch fiber (OPEBF). The fiber in the fresh fruit or palm-pressed fiber is also referred to as OPMF [42]. Both fibers are extracted through different techniques.

EFB fiber is extracted through any one of the retting processes, namely, mechanical retting (hammering), chemical retting (boiling with chemicals), dew/vapor/steam retting and microbial/water retting; water retting being the common practice [15]. Up to about 73% of fibers can be obtained from OPF source [48]; either bunch (60%) or mesocarp (73%) [49]. Mechanical retting is also common due to its relative environmental friendliness as compared to other processes [15].

A device designed by Jayashree et al. [50] to mechanically ret fiber from bunch is also able to decorticate oil palm empty fruit fiber, disaggregate the pith, sort and group the fiber into distinctive groups. Since palm oil is actually embedded in palm fruits, the fruit must undergo further processing before mesocarp fiber is released [42]. The fruit, therefore, first undergoes digestion or pounding process. Industrially, a steam-heated cylindrical vessel fitted with a central rotating shaft surrounded with a number of beaters (stirrers) fulfills the digestion process. Within the digester, content is pressed dry to squeeze out palm oil. The wet method applies hot water after digestion, which leaches out palm oil, allowing it to settle on the surface of liquid solution through the process of sedimentation. While by-products of dry process are moisture OPMF and palm nuts, wet process consists of effluent, OPMF and palm nuts. In any of the processes, OPMF is separated from other content materials [42] if they are to be put in further use as waste materials. The wet and dry processes of releasing oil after digestion may not be mutually exclusive.

The OPF has an empty middle portion with empty cylindrical porous frame [51]. Resembling coir fibers, OPF is tough and hard [45, 52]. Average pore diameter is 0.07 μm and cell structure width 0.128 mm. The porous surface structure is responsible for better interlocking characteristics when used in matrix polymer composite formation [45]. It is believed that each fiber is a collection of tiny individual fibers, thus constituting a bundle of other tiny fibers [3, 17, 53]. Inside the vascular bundles are starch granules, while silica bodies surround the fiber strands. Circular craters that are steadily distributed on the fiber surfaces are attached by the silica bodies [54].

Average granular starch diameter, circular crater diameter and silica body diameter size may be around 0.115, 0.112 and 0.088 mm, respectively [55]; thus when the silica bodies are removed by either mechanical or chemical means or otherwise the empty craters facilitate better interlock between fiber and matrix in composite formation, resulting in enhanced fiber–matrix interfacial adhesive bonding. Though the characteristic morphology of OPF inure to it the benefit of enhanced hardness and toughness [56] and high cellulose content [14], the occupation of hydroxyl group interferes with interfacial bonding within the hydrophobic polymer matrix composites [3]. Cellulose content may vary between 14% and 20% [57].

The presence of hydroxyl and carboxyl groups in cellulose introduces incompatibility with nonpolar plastics [58], leading to aggregate formation during extrusion; difficulty to homogenize; display of limited tress transfer and poor adhesion; and display of fiber pullouts in fracture surfaces. This may negatively affect the mechanical and physical properties of composites [15]. Thus the poor performance of fiber–polymer composites in terms of physical and mechanical characteristics can be attributed to porous tubular structure of fibers and high cellulose content present. The presence of palm oil surrounding fibers may also interfere fiber–matrix compatibility and also reduce coupling efficiency of coupling agents [59]. Nevertheless, these limitations can be reduced by improving fiber properties through surface modification to enhance interfacial bonding between fiber and polymer matrix during composite fabrication. Tables 4.1 and 4.2 show the chemical composition and physicomechanical properties of OPF, respectively, as identified by various researchers since 1991.

4.3 Oil palm fiber composites

Many factors have contributed to the shift to biofiber-related technology. First, scarcity of natural resources such as shortage of landfill space and environmental pollution including greenhouse emissions and residue run off have been a motive behind global interest in biobased materials as alternative means in materials

Table 4.1: Chemcal composition of OPF.

Constituents (%)	Range	References
Cellulose (%)	42.7–65	Then et al. [49], Khalil et al. [40], Law et al. [54], Hill and Khalil [13], Sreekala et al. [45], Khoo and Lee (1991)
Lignin (%)	13.2–25.31	Then et al. [49], Hill and Khalil (2008b); Ismail et al. (1997); Sreekala et al. [45], Wiijosentono et al. [48], Abubakar et al. [60], Kahlil et al. (2007a); Rozman et al. [61]
Hemicellulose(%)	17.1–33.5	Hill and Khalil (2000 b): Khalil et al. (1997); Rozman et al. (2007)
Holocellulose (%)	68.3–86.3	Khoo and Lee (1991); Ismail et al. (1997); Law and Jiang (2001). Abubakar et al. [60], Law et al. [54], Rozman et al. [61], Khalid et al. (2006 a); Khalil et al. (2008 b)
Ash content (%)	3.0–6.04	Then et al. [49], Ismail et al. (1997): Sreekala et al. [45], Law & Jiang (2001); Wirjosentono et al. [48], Abubakar et al. [60], Khalil et al., (2007 a). Law et al. [54], Rozman et al. [61], Khalil et al. (2008 b)
Extractives in hot water (100 °C)(%)	2.8–14.79	Khoo and Lee (1991): Law and Jiang (2001); Law et al. [54], Khalid et al. (2008 a).
Solubles in cold water (%)	8–11.46	Sreekala et al. (1997); Wirjosentono et al. [48]
Alkali soluble (%)	14.5–31.17	Khoo and Lee (1991); Ismail et al. (1997); Sreekala et al. [45], Law and Jiang (2001); Wirjosentono et al. (2004); Law et al. [54]
Alpha-cellulose (%)	41.9–60.6	Ismail et al. (1997); Abubakar et al. [60], Khalid et al. [62]
Alcohol–benzene solubility	2.7–12	Ismail et al., (1997); Sreekala et al. [45], Wirjoentono et al., [48], Khalid et al. [62]
Pentosan (%)	17.8–20.3	Ismail et al., (1997); Khalil et al., (2007a)
Arabinose (%)	2.5	Law et al. [54].
Xylose (%)	33.1–24.01	Yusof et al. (2009)
Mannose (%)	1.3	
Galactose (%)	1.0	
Glucose (%)	66.4–42.82	
Silica (EDAX) (%)	1.8	
Copper (g/g)	0.8	
Calcium (g/g)	2.8	
Manganese (g/g)	7.4	
Iron (g/g)	10.0	
Sodium (g/g)	11.0	

Source: Shinoj et al. [3], Then et al. [49], Yusof et al. (2009).

Table 4.2: Physicomechanical properties of OPF.

Property	Range	Authors
Diameter (μm)	150–300	Sreekala and Thomas (2003); Bismarck et al., (2005), Jacob et al., (2006b)
Microfibrillar angle (º)	46	Bismarck et al., (2005)
Density (g/cm)	0.7–1.55	Sreekala and Thomas (2003); Kalam et al. (2005); Khalil et al. (2007a); Rao (2007).
Tensile strength (MPa)	50–400	Sreekala et al. (2004); Bismarck et al. (2005) Kalam et al. (2005); Abubarkar et al. (2006)
Young's modulus (GPa)	0.57–9	Yousif and Tayeb (2008) Sreekala et al. (2004); Bismarck et al. (2005) Abubarkar et al. (2006); Yousif and Tayeb, (2008); Jacob et al. (2006b); Khalil et al. (2007a);
Elongation at break (%)	4–18	Then et al. (2013); Bismarck et al. (2005); Bismarck et al. (2005); Abubakar et al. (2006); Jacob et al. (2006 b); Kahlil et al. (2007 a); Yousif and Tayeb (2007)
Tensile strain	13.71	Rao and Rao (2007)
Length-weighted fiber length (mm)	0.99	Law et al. (2007)
Cell wall thickness (μm)	1.3–3.38	Chang et al. (2017)
Fiber coarseness (ma/m)	1.37	Shinoj et al. (2011)
Fineness (<0.2 mm) (%) Rigidity index $(T/D)^3 \times 10^4$	27.6 55.43	

Source: Shinoj et al. [3], Chang et al. [55], Then et al. [49].

technology [63, 64]. Second, recent introduction of domestic and international legislations have further engaged industry players to divert attention from traditional and nonrenewable materials and resources to new renewable resourced materials [65]. Third, with reference to economic and environmental cost of waste disposal, it is pertinent that society and industry players seek different technologies to deal with waste materials (Väisänen et al., 2016). In addition to these, the introduction of circular economy and life-cycle assessment of manufactured products with associated economic philosophies such as green economy, bioeconomy, biobased society and ultragreen development are presently dictating long-term strategies of industry players [66]. Finally, the old and traditional methods of disposing waste has become unattractive as a result of stern, compelling and sometimes captivating legislations in the form of tax waivers and tax penalties (Väisänen et al., 2017).

Other factors, including the following, have also been identified to contribute to the shift from traditional materials to the current interest and use of composite materials, particularly natural fiber–plastic composite materials. First, there is unreliability of crude oil prices and the competition of petroleum-based materials, of which plastics are generally manufactured from the US Energy Information Administration (IEA), 2013 [67]. Second, general society has found interest in alternative material products as consumer products instead of conventional ones (Väisänen et al., 2016). Third, with the continuous expansion of demand and production quantum of natural fiber-based composites, there has arisen political interest in terms of production and also demand for measurable supply of raw materials with regard to quantity and above all, quality (Väisänen et al., 2016).

Finally, the availability and cost of traditional materials are being questioned as industries are now encouraged to convert production technologies to natural fiber-based composites [67]. It is well noting that the crave for corporate image building and its protection and sustenance coupled with international and global policies, environmental awareness as well as financial drivers have in one way or the other contributed and supported the utilization of natural fiber in both primary and secondary applications [68]. Within industrial and consumer circles, the need for availability in terms of quantity and quality of product cannot be overemphasized. There is also the need for adequate and quality supply of raw materials to feed industries at the primary, secondary and tertiary levels of natural fiber-based commodities. Logistics and technological as well as product competiveness also need to be addressed. Thus, quality raw materials, whether refined or otherwise, should be an all-important production parameter as far as natural fiber utilization is concerned in the development of composites.

The use of OPF in development of natural fiber–plastic composites has come into sharp focus in recent times, particularly, in countries where the palm tree is found relatively abundant. Many research works have therefore been done of late with the purpose of developing thermoset and thermoplastic composites filled with OPF [3, 49]. Composite research works have been related to characterization of physical, thermal, electrical, mechanical and biodegradation properties. Characteristic features mostly found in studies include creep behavior as a predictor of durable mechanical behavior, stress relaxation, directional stability concerning loading elements as well as detention of damping forces [69]. Structural and load-bearing studies with reference to tensile, flexural, impact and static characteristics (mechanical) have also been pursued. Studies relating electrical properties were based on composite utilizations relating electrical-electronic component applications, including insulations. Crystallization kinetics in higher and lower temperature utilizations have also been researched among OPF composite materials. Research work concerning water absorption and swelling characteristics of components with reference to wastewater treatment, packaging and handling as well as building construction has equally been carried out.

Composite resistance to fungal attack as a measure of resistance to biodegradation toward indoor and outdoor conditions during the life cycle of oil palm–fiber–plastic composites is also not left out in various studies [70]. It must be emphasized that composite properties may depend on the method of testing employed and selection of matrix in terms of desired composite properties. The characteristics of end product composite also depend on variables such as fiber size, fiber–plastic weight ratio, fiber–plastic volume ratio, fiber orientation, fiber surface treatment and type of plastic matrix material. Plastic materials considered in this chapter include PES, EP and PU. The rest are phenol formaldehyde, PVC, PP, NR and PE. Tables 4.3 and 4.4 display the respective standard methods employed by various authors to determine mechanical as well as physical, thermal and biodegradation characteristics of OPF–polymer blends.

Table 4.3: Common standard methods for conducting mechanical testing of OPF composites.

Property	Standards	Authors
Tensile strength	ASTM D 638	Ghazilan et al. (2017); Olusunmade et al. (2016); Owulonu and Egwe (2012); Yusoff and Salit (2010); Then et al. (2013); Mohammed et al. (2015); Eng et al. (2014); Chollakup et al (2013); Karina et al. (2008); Bakar and Baharulrazi (2008): Hariharan and Khalil 92005) Khalili et al. (2008 a); Sreekala et al. (2002 a); Wirjosentono et al. (2004).
Tear strength	ASTM 624 BS 903-A3	Jacob et al. (2008); Jacob et al. (2004) Ismail et al. (1997)
Flexural strength	ASTMD 790	Rozamn et al. (2007): Wijosentono et al. (2004); Abubakar et al. (2007); Abubakar et al (2007): Abubakar et al. (2005); Sreekala et al. (2002a); Then et al. (2013); En et al. 92014); Karina et al. (2008); Yusoff and Salit (2010).
Impact strength	ASTMD256	Olusunmade et al. (2016); Then et al. 92013); Eng et al. (2014); Badri et al. (2006); Bakar and Baharulrazi (2008); Hariharan and Khalil (2005); Khalil et al. 92007 a): Khalil et al. (2007 b);
Hardness	ASTM 2240 ASTM785	Badri et al. (2006); Ismail et al. (1997); John et al. (2008); Raju et al. (2008); Sereekala et al. (2002 a). Ewulonu and Egwe (2012); Olusunmade et al. (2016)

Source: Shinoj et al. [3], Ghazilan et al (2017), Olusunmade et al. [1]., Owulonu and Egwe [71], Yusoff and Salit [72], Then et al. [49], Mohammed et al. (2015), Eng et al. (2014), Chollakup et al. [73].

Table 4.4: Common standard methods for testing physical, thermal and degradation properties of OPF composites.

Property	Standard	Authors
Flame propagation	ASTMD 4804	Ewulonu and Igwe (2012)
Water absorption	ASTMD570	Amin and Badri (2007); Bakar and Baharulrazi (2008); Badri et al. (2006)
Density	ASTMD792 BS4370-1	Ewulonu and Igwe (2012); Sreekala et al. (2002a); Badri et al. (2005)
DSC	ASTMD3418-82 ISO 1180	Bakar and Baharulrazi (2008); Tesfamichael and Nor (2014)
Degradation	BS: EN150 486 1997	Hill and Khalil (2000)

Source: Shinoj et al. [3] and Ewulonu and Igwe [71].

4.4 OPF–NR blends

Latex may be obtained through rubber tree sap, while NR is a product of latex. Rubber is used to produce several products, including industrial products and consumer household items. Vehicular tire and tube products take the chunk of items produced from NR. Other products classified as General Rubber Goods take the remaining chunk. For the past five decades or so, several research works and studies have been conducted relating OPF–NR composites and their characterization. Review of literature from 1998 has been presented in this chapter with particular reference to mechanical, thermal, water-absorption and electrical properties.

Investigating the effect of fiber size in terms of length, Joseph et al. [41] identified a critical length (l_c) lesser or greater than which the characteristics of composite differs. At I_c, load transmission from NR to filler reached greatest. When fiber length was less than I_c, there was composite failure even when load is low. This was due to bond separation between stressed fiber and matrix. Tensile strength of blends reached maximum when the length of fibrils reached 6 mm but decreased at increasing fibril length. The reason was because at greater fiber length, OPFs were more entangled. The study also observed that the orientation of fibers in composite has an effect on tensile properties. For instance, transverse orientation of fibers yielded 17.5 MPa, while longitudinal orientation yielded 19.2 MPa. Break elongation for transverse orientation was higher (940%) while that of longitudinal orientation was 108.2%. It was explained that fibers transversely oriented were perpendicularly aligned to load direction, thus preventing them in the participation of stress transfer. On the other hand, longitudinally oriented fibers were aligned parallel to load, thereby causing uniform load transmission.

The impact of filler with reference to weight was also investigated by Joseph et al. (2006a; 2006b) and Jacob et al. (2004). While Jacob et al. (2004) worked on OPF–sisal fiber–NR (OPF–sisal fiber–NR or oil palm–sisal–NR) composites, Joseph et al. [41] studied OPF–NR composites. The studies identified that when OPF is bonded with NR, tensile strength and elongation at break decreased. The reason was attributed to higher strength of NR, due to strain-induced crystallization, reducing as a result of disruption of regular arrangement of the crystals. Comparatively, the tensile strength of pure gum was greater with reference to that of oil palm–sisal–NF hybrid blend [56]. The studies also showed that as fiber content increases, stress relaxation rate of OPF–sisal fiber–NR hybrid composites decreases as compared to pure gum. The presence of fiber in the fiber–rubber interface obstructs the relaxation contributed by the pure rubber, thus decreasing the relaxation rate of the composite. At variable temperatures, it was revealed that storage modulus of oil palm–sisal–NR blends elevated with increased fiber content. The NR phase ensures more material flexibility, which contributes to low stiffness and consequent low storage modulus. Increased storage modulus was brought about as a result of greater stress transfer and increased composite stiffness at the composite material interface when more fiber was added. The studies also revealed that loss modulus elevated with filler weight to 756 MPa maximum at 50 phr, while 415 MPa was recorded for pure gum. When fibers were added to rubber, the damping characteristics of NR decreased as a result of flexibility reduction due to restriction in degrees of molecular motion.

Jacob et al. [74] established the generic assertion that blending with vegetative fiber increases polymer matrices' resistance to heat. It was also found that thermal stability increases with increase in fiber content as a result of decrease in activation energy of oil palm–sisal fiber–NR blends. With reference to aqua-absorption properties, Jacob et al. [75] observed that the proportion of water absorbed by OPF–NR blends elevated according to the enhanced fiber loading as a result of water affinity behavior of natural fibers. Due to the viscoelastic characteristic of NR and the presence of microcracks when OPF was added, the aqua-absorption feature of NR with Fickian characteristics was transformed to non-Fickian behavior. Another finding was the fact that water transmission into OPF–sisal fiber–NR hybrid biocomposite was higher than OPF–NR composite. This was explained by the fact that sisal is more cellulosic than OPF to the tune of 23% while lignin content of OPF is higher than sisal by 19%. It was concluded that cellulose is hydrophilic while lignin is hydrophobic and this accounts for the comparative difference between OPF–sisal fiber–NR and OPF–NR composite materials.

Electrical properties of OPF composites were also explored by Jacob et al. [76]. The polarizability of composites increased when its dielectric constant improved. Total polarizability is the cumulative effects of atomic, electronic and orientation of the constituents of the composite material. In an OPF–sisal fiber–NR composite, there was high polarization as a result of increased nonpolar lignocellulose fibers amidst polar NR. For these reasons, dielectric constant of composite was found to be higher as fiber loading increased at different frequency levels. The study further found that loss tangent (tan δ) at 10 Hz elevated by 0.033 from 0.002 when 50 phr OPF–sisal fiber was combined with NR [76].

It was explained that fiber addition thus results in higher quantity of dissipation that is usually characterized by relaxations in amorphous phases. The reasons were twofold: first, cellulosic content of fiber makes it hydrophilic, thus enhancing the flow of current through the amorphous polar region. Second, there was loss of crystallization capability as a result of the disruption of regular grouping of rubber molecules due to the presence of fibers in the NR matrix. The dissipation factor of oil palm–sisal–natural fiber composite, therefore, enhanced as fibril component was elevated. Furthermore, the study found that at 10 kHz NR volume resistivity lessened by 400×10^4 Ωm when 50 phr OPF–sisal fiber was combined with NR [76]. Higher volume fractions ensured regularly uniform arrangements and facilitate smooth flow of current due to the presence of polar groups in OPF and sisal fiber. Higher fiber volume fractions further ensure organized, regular and more oriented fibers, which also promote smooth current flow. The presence of moisture in the amorphous region of fiber and the availability of crystalline region in the polymer were combined to enhance reduction of volume resistivity of oil palm–sisal–NR polymer, thus resulting in increase in volume conductivity.

4.5 OPF–PP blends

Polypropylene (PP) is a thermoplastic polymer produced through polymerization of the monomer propylene. It is resistant to chemicals such as basic and acidic chemicals. PP is therefore applicable for the manufacture of several consumer products such as automotive parts; handling and packaging products; laboratory equipment and items; and office and stationery accoutrements. It is amenable to production techniques such as injection and extrusion molding. As a result of its low cost, distinguished mechanical characteristics and easy ability to be processed was the first synthetic polymer to gain industrial as well as commercial accessibility.

The tensile property of PP has been found to reduce upon OPF also referred to as empty bunch fiber loading in OPF–PP composites [59, 77]. By increasing the fiber content of OPF from 10% to 30% and 40%, tensile strength of OPF–PP blend lessened from 19.3 to 19 MPa, and 12 and 7 MPa, respectively, with corresponding increase of tensile modulus from 300 to 5,600 MPa, and 65 and 725 MPa, respectively. This behavior was attributed to the reduction of stress transfer through matrix as a result of incompatibility of PP and OPF molecules as well as size irregularity of OPFs [2, 3, 77].

The mechanical characteristics of EFB–PP and EFB-derived cellulose-PP were also investigated by Khalid et al. [78] and Rozman et al. [59]. The study revealed that OPF loading in PP matrix reduced the tensile strength. This was due to obstruction of the fiber to transfer stress in the direction of applied force resulting from the absence of attractive molecular forces between plastic and filler. However, inclusion of oil-palm-based cellulose improved tensile strength. Nevertheless, tensile strengths were relatively lower at lower cellulose proportions. When 40% weight of cellulose was included, tensile strength was found to be the same as homogeneous PP. While the flexural modulus of composite enhanced with increase in fiber weight and that of cellulose-impregnated blend was higher than fiber composites. Whereas flexural modulus of OPF blend of 50% filler composition improved to a maximum of 2,750 MPa, 50% cellulose enhanced flexural modulus by 2,500 MPa. Nevertheless, increasing the proportion of OPF caused reduction in flexural strength. Wirjosentono et al. [48] through reactive processing obtained 21.4 MPa maximum tensile strength when 20% weight proportion of fiber was filled with PP though break elongation decreased with increase in fiber loading.

Rozman et al. [79] also recorded modulus of flexure, flexure and toughness flexure of the OPF–PP blends to be 2.13 GPa, 37 MPa and 47 kPa, respectively. The impact strength of PP reduced from 7 J/mm, when 10% fiber weight was applied; but constantly elevated to 37 J/mm when up to 50% fiber was included [79]. It was also observed that the same impact strength was exhibited by pure PP when 50% OPF loading was applied. Low aspect ratio and irregular shape of OPF may contribute to enhanced stress transmission from PP to the matrix [79]. Increasing the percentage of OPF in OPF–glass–PP composite reduces flexural properties [77]. Thus when glass fiber was fully replaced by OPF, flexural modulus, flexure and toughness flexure lessened from 20 MPa, 37 MPa and 3.75 GPa to 25 kPa, 2.75 GPa and 13 kPa, respectively [59]. Similarly, at 30% fiber loading, OPF–glass fiber–PP composite tensile strength, tensile modulus and tensile toughness reduced from 25 MPa, 850 MPa and 180 kPa to 13 MPa, 610 MPa and 105 kPa, respectively, after fully replacing glass fiber with OPF.

In terms of water absorption behavior, Rozman et al. [79] record that elevated percentage of OPF yields higher aqua-absorption characteristics as a result of higher proportion of hemicellulose, lignin and cellulose. These bodies possess hydroxyl group, which is polar in nature, and therefore has higher affinity to water due to the presence of hydrogen bonds. When OPF–PP composites were immersed in water for 6 days, water absorption ranged from 4.5% to 75% while percent swelling thickness was between 2.3% and 7%. The work of Hill and Kahlil [13] also exhibited the effect of weathering or degradation on OPF–PP composites. It was observed that flexural stress and flexural modulus lessened by 24.71 MPa and 37.75 GPa, respectively, when composites were engaged in 12 months soil exposure.

4.6 OPF–PU blends

Studies have shown that the addition of OPF in PU matrix is a good reinforcer. PP contains a minimum two each of isocyanate functional group and hydroxyl functional groups. These two monomers are chemically combined in the presence of a catalyst to produce PU. In the construction and building industry, PU is used to produce door frames, columns, window headers and ceiling panels, among others. It is also used to produce electronic components, carpets, microcellular foam seal and gaskets, high resiliency flexible foam seats, vehicle automatic suspension bushings and rigid foam insulation panels [3].

Studies have also shown that increase in fiber loading increases the tensile strength and tensile modulus of composites. Increase in tensile strength was as a result of fiber acting as a reactive component and filler. This is caused by increase in isocyanate and lignin reaction when a three-dimensional networks and crosslinks are formed as a result. Increase in strength modulus results increase in composite hardness. The presence of hydroxyl group in fiber donates electrons while carboxyl group in PU accepts; this generates physical cross-links in the composite material, thus producing additional adhesive forces between fiber and PU interface [61, 79].

OPF-filled PU composites were found to decrease in flexural strength when smaller fiber size was employed. Tensile strength was also found to decrease when fiber size reduced [79, 80]. However, compressive stress increased when smaller fiber particles were filled in the composite [81]. This was explained by the fact that small size fibers are able to fit in better within the matrix. Additionally, stress-impact propagation might have also been obstructed due to increased surface area of powdered fiber used in the fabrication.

It has also been found that OPF–PU composites have weak water absorption characteristics. For instance, Badri et al. [81] observed that OPF–PU–OPF blend with 75% fibril weight composition ingested 55% water, while 65% fibril composition ingested 48% within 24 h of composite water interaction. Three reasons were attributable: insufficient encirclement of OPF by matrix, weak clustering of OPF and fragile fiber–matrix bonding. The authors also observed that larger sizes of OPF in OPF-filled PU matrix exhibited increased deformation as a result of cellular structure degeneration of PU foam. In spite of this, the dimensional stability of composite with 45–56 μm OPF, though larger in size, was within allowable limits.

4.7 OPF–PVC blends

Homogeneous PVC is a brittle white solid soluble in alcohol. As a thermoplastic polymer, it places third to the commonly used plastic, besides PE and PP. Basically, PVC can be produced in two forms, flexible and rigid from vinyl chloride monomers. Rigid

PVC is used in the construction of profile applications for doors and windows as well as pipes; manufacture of cards (ATM, membership, identity or bank cards), bottles and nonfood packaging. In flexible form, it is utilized in the production of canvas, rubber and plumbing products. Other consumer products from flexible PVC include electrical cable insulation, flooring and phonographic products. It is resistant to biological, acidic and basic substance attack.

Bakar and Baharulrazi [82], Abubakar Hassan and Yusof [83] and Abubakar, Hassan and Yusof [60] did some work on the mechanical characteristics of OPF–PVC blends employing varying fibril compositions. Findings were indicated as follows: as fiber loading decreased and tensile strength increased; reduction in tensile strength was caused due to collapsed longitudinal edges of truncated fibers. Deformation of tensile strength was caused by activated fractures within PVC. Other observations include the fact that irregular shape and consequent dispersion challenges (ogglomeration of filler/assemblage) were responsible for OPF ability to bridge stress transmission from the matrix. At high fiber loading, impact strength was observed to reduce as a result of inability of OPF to transmit stress via shear yielding before fracture. Flexural strength also reduced when percentage fiber was increased due to incapability of fiber to bridge transfer of stress from the matrix as well as dispersion challenges (agglomeration of filler).

The characterization of OPF, PVC and epoxide NR composite was established by Raju et al. [15]. Tensile strength, ultimate tensile strength and impact strength reduced with increasing OPF loading up to 30%. Elongation was also reduced. However, flexural and tensile moduli were improved. There was marginal improvement of hardness when fiber weight was above 5%. Reduction in characteristics such as tensile strength and impact strength was caused by filler particle agglomeration that acts as a foreign domain body to transmit stress to the matrix. Increase in tensile and flexural characteristics of composite were due to reduction in polymer chain elastic behavior when fiber was added, with consequent rigid composite characteristics. Fundamentally, the presence of OPF did not impact on heat resistance of the composite. However, the glass transition temperature (T_g) of the blends improved when fibril weight was enhanced.

4.8 OPF–PE blends

PES as a plastic consists of one or more ester units within the primary functional macromolecule. PESs may be synthetic or natural. Commonly referred PES is the material polyethylene terephthalate (PET). PET may be thermosetting plastic or thermoplastic. However, the most used PESs are thermoplastics. PESs are used in the manufacture of textiles, bottles, canvas, film insulation for wires, insulating tapes, dielectric film for capacitors, among others.

Studies have attested that OPF–PES composites exhibit lower flexural strength than pure PES [44]. Chin [84] also attested that by increasing OPF loading up to 30%, tensile strength of composite decreased while tensile modulus increased to about 56.7%. A study by Mahjoub et al. [77] exhibited that tensile strength of OPF–PES composites is less than pure PES; tensile modules of OPF–PES composites are higher than pure PES; there was reduction of elongation at break of OPF–PES composites compared to pure PES; there was decrease in ultimate tensile strength as a result of reduction of strain energy of the composite; and OPF–PES composite had higher flexural modulus but reduced flexural strength than pure PES. The study therefore concluded that, generally, addition of OPF to PES has negative effect on mechanical behavior of PES matrix.

Then et al. [49] studying comparative characteristics of OPEBF and OPMF also observed flexural and tensile moduli increase when fibers were filled in PES matrix. The opposite was however observed for impact, tensile, flexural strengths and elongation at break. The study also observed that OPMF–polymer composites possessed better tensile, impact and flexural strengths as well as tensile and flexural moduli and elongation at break than OPEBF–PES composites. The study further observed that reduction in mechanical characteristics modified as fibril weight was raised from 10% to 70% by weight, after which the trend changed. According to Lee et al. [85], this characteristic behavior is due to of blocking of segmental migration of plastic intermolecular linkages, which in turn limits deformability of matrix phase, rendering composites stiffer; thus, improving flexural and tensile moduli but reducing elongation at break and strength behaviors of composites. This is due to the fact that polar nature of OPF and nonpolar behavior of PES might result in lack of interfacial interaction, thus interrupting stress transfer from fiber to matrix [77].

OPF–PES composites have better abrasion properties than pure PES. Reinforcement of OPF filler might reduce weight loss on abrasion by 50–60% when compared with unfilled PES. The friction coefficient of pure PES might also be higher than OPF–PES composite by about 23%. Furthermore, the tensile strength of OPF and glass fiber-filled PES hybrid blends enhanced continuously to 45% maximum loading of OPF. Thus, in applications where high load bearing forces may not be expected, OPF could replace glass fiber. Recognizing the fact that density of PES (1.202 g/cm^3) is higher than that of OPF powder (1.138 g/cm^3), adding 55% OPF in matrix lessened the specific gravity of PES to 1.17 g/cm^3. Similarly, since the density of glass fiber (2.6 g/cm^3) is higher than that of OPF, replacing glass fiber with 40% (and more) of OPF reduced density of PES from 1.15 to 1.11 g/cm^3 of OPF–PES composite [13, 53, 86].

Studying the water ingestion behavior of OPF–glass fiber–PE composite, Khalil et al. [86] observed that the absorption capacity of pure PES (1%) increases to 9% when 45% OPF was added. When 70% PES and 30% glass fiber was added, water absorption capacity reduced to 6%. When 12% glass was incorporated with pure PES and placed in boiling water for 120 min, swelling capacity improved by 0.16% [44]. In addition to this, when 70% of OPF fiber fraction was incorporated swelling capacity increased to 2.03%, while 2.46% swelling capacity was obtained when OPF fiber content was further increased to 100%. When immersed in water, OPF–PES composite size increased linearly with the square root of time and gradually decreased till complete saturated condition or equilibrium was reached; known as the Fickian characteristics (Khalil et al. 2008). This behavior was explained by the fact that immersion results in swelling of fibers, which cause cracks to be formed in the matrix and consequently creating empty paths for water molecules to easily move into the composite fabric.

With reference to composite degradation or weathering, Khalil and Ismail [70] and Hill and Khalil (2001) found that when OPF–PES composites are buried under soil conditions, for 3, 6 and 12 months, tensile strength reduced by 8%, 17% and 35%, respectively. Similarly, the respective impact strengths reduced by 6%, 18% and 43% within the same periods of time. Tensile modulus, tensile stress and elongation at break also lessened from 3.29 GPa, 35.1 GPa, and 3.75 GPa to 2.32 MPa, 34.6 MPa and 2.48 MPa, respectively, when buried for 12 months.

4.9 OPF–PS blend

Polysterene (PS), as aromatic plastic, is produced through polymerization of styrene. PS is one of the most widely used plastics for several reasons. It is inexpensive in terms of unit weight, transparent, hard and brittle. It is resistant to water vapor and oxygen penetration but has relatively low melting point and comparatively resistant to biodegradation. As a thermoplastic, its applications include plastic models, digital versatile disk and compact disk production. In the foam form, it is used for packaging, foam drinking cups, electrical insulation materials, building insulation for walls, floor and ceiling paneling, and low load-bearing building structures. Though it is one of the most commonly applied plastic within the consumer market studies relating to its OPF composites are sadly limited in literature.

A study by Zakaria and Poh [87] on flexural characteristics of OPF–PS with 10% fibril loading for 300–500 µm size fibrils obtained 1,665.4 MPa, 2,685.8 MPa and 0.027 mm for maximum stress, modulus of elasticity and Young's modulus, respectively. Modulus of flexure increased with fiber content up to 30%, while flexural strength and maximum stress reduced. Reduction in strain was attributed to irregularity in fiber shape resulting in stress transfer obstructions from fiber to matrix. The study identified that fiber size less than 300 µm had no effect on flexural characteristics. However, benzoylation treatment enhanced interfacial adhesion and hydrophobic characteristics, thus improving flexural properties of the composites.

4.10 OPF–EP blends

Many studies avail for OPF and EP blends. EP resins are polymers that contain the epoxide groups. EP resins may react with themselves or coreactant to form thermosetting polymers with better mechanical properties as well as thermal and chemical resistance. Applications of EP include structural adhesives, paint brush manufacturing, metal coatings, electronic and electrical components, high tension electrical insulators and fiber-reinforced plastic composites. EP or polyepoxide has been found in many OPF–plastic fabrications.

A study by Yusoff and Salit [72] indicated that tensile strength of pure EP was higher than OPF-reinforced composites though the difference was not much when OPF loading varied from 5% to 20% by volume. The tensile strength varied by 21% between the lowest and highest average tensile strengths. Load transfer was hindered as a result of chopped nature of fiber, which randomized the distribution of fiber in the matrix. Young's modulus was also found to be at peak at 5% volume of OPF, decreasing with increasing loading. The study attributed this characteristic to be due to fiber pullout as loading increased. Fiber pullout created voids and small gaps due to insufficient wettability or bonding between fiber and matrix during composite fabrication. Other factors that influenced the mechanical properties include fiber content aspect ratio of fiber length, orientation and distribution of fibers, intrinsic characteristics of both fibers and matrix, and fiber– matrix adhesion which affects load transfer within the composite.

Yusoff and Salit [72] also indicated that the flexural strength of OPF–EP blends decreased as fibril volume fraction improved. The highest flexural strength was 51 MPa at 10% volume fraction of OPF, while the lowest was 40.0 MPa when volume fraction was 5% by volume. Flexural strength of composite was also lower than that of pure EP. The effect of fiber content on flexural strength of EP was also indicated to be insignificant. On the other hand, flexural moduli tend to decrease when fiber content was increased. This was indicated to be due to weak fiber–matrix bonding. Mahjoub et al. [77] obtained similar results; when 55% of EP matrix was replaced by OPF by volume, both tensile strength and tensile moduli reduced from 62.49 MPa and 1.13 GPa to 46 MPa and 1.02 GPa, respectively. Ghazilan et al. (2017) have also confirmed that tensile strength and tensile modulus decreased when fiber loading is increased within EP matrix. The study showed that at 20% fiber volume, tensile strength and tensile modulus decreased by 61% and 38%, respectively, relative to pure EP.

The effect of OPF loading decreased the tensile strength of glass fiber–EP blend by 87 MPa by replacing mineral fiber (glass) with OPF [16]. This was caused by weak interfacial bond between OPF and EP matrix. This in turn caused reduction in stress transfer between the two materials. It was explained by the fact that strong pectin interface between individual fibers hindered loading abilities of fiber on matrix yielding restricted loading or no loading at all. The strength property of hybrid composite was therefore intermediate in nature.

Bakar et al. [88] using a 7-m beam configuration observed that OPF–EP composite with 10% fiber loading could yield a maximum deflection of 0.2 mm when supporting 200 kg load. Comparatively, OPF is limited in capacity as an enforcer in relation to carbon fiber in EP matrix. In other words, the contribution of OPF to fatigue strength in EP matrix was insignificant since addition of OPF rather reduced the ultimate strength and fatigue resistance of matrix. This was substantiated by the fact that tensile strength and ultimate tensile strength of OPF–EP composite lessened by 1.68 MPa and 184.9 MPa when fiber loading was increased by 20% from 35% (Kalem et al, 2005). The study also indicated that OPF–EP composites could be employed in design and construction of structures that may support moderate loads, such as short span bridges and housing beams. Hybrid OPF–glass fiber–EP composite exhibited highest elongation at break followed by glass fiber–EP blend and OPF–EP blend in that order. Since the tensile modulus of glass fiber is higher than that of OPF, the introduction of glass fiber in matrix introduces relatively increased stiffness into the matrix. It was also revealed that load was shared between the two reinforcement materials, thus introducing those distinctive behaviors of the three composites. The glass fiber was found to first receive the load and then redistributed excess to OPF. With reference to impact loading, glass fiber–EP composite showed higher impact strength (107 kJ/m^2) than OPF–EP blends (18 kJ/m^2).

4.11 OPF–PE blends

PE or polymethane, also referred to as polythene, is the most globally used plastic. Common types of PE include high-density PE (HDPE), low-density PE (LDPE) and linear low-density polyethylene (LLDPE). HDPE is used for packaging products such as garbage containers, water pipes and detergent bottles. LDPE is used in the manufacture of plastic bags and film wraps as well as rigid containers, food bags and hosing and tubing. A few studies have been attended to composite formation and characterization with regard to OPF.

The dynamic mechanical properties of oil palm fiber (OPEFB) and LLDPE composite was studied by Shinoj et al. [3]. Corresponding fiber–matrix interactions were also studied. The study observed that storage modulus of LLDPE increased with increase in fiber content. Loss modulus also enhanced with the increase in fiber loading with the exception of composite with 10% fiber loading. The study also observed that with 40% fiber loading, glass transition of pure LLDPE increased from −145 to −128 °C. It was further observed that loss modulus peak of OPF–LLDPE composite increased to a maximum at 10% fiber content. The study found tan δ peak values decreasing upon fiber loading, with higher tan δ values exhibited by composites of 177–425 μm size proportion.

Between temperature range −40 and 60 °C, the minimum tan δ values have been recorded for lesser dimensioned fibrils. The study also found that interfacial strength indicator (β) increased with the increase in frequency. Bond strength decreased with fiber size from 75–177 µm, 425–840 µm to 177–425 µm in that order. It was further found that activation energy of pure LLDPE was lower than that of composites except when fiber loading was 10%. Activation energy was observed to increase for composites with 75–177 µm size fraction, 177–425 µm size fraction and 425–840 µm with respective values of 68.1 and 80.7 kJ/mol. It furthermore determined that predicted storage modulus and damping variables, employing theoretical equations, were in agreement with experimental values. The effect of surface treatment employing alkaline solution also had an impact on the dynamic mechanical characteristics of the blends.

Some properties of OPEFB fiber–HDPE composite with maleic anhydride as compatibilizer were analyzed by Ewulonu and Igwe [71]. In general, the study showed that maleated HDPE composites had significantly better properties with reference to hardness, tensile strength, elongation at break, specific gravity, flame resistance and water sorption. The study identified that tensile strength and elongation at break showed a decrease as fiber loading increased. However, specific gravity and hardness increased with increase in OPEFB fiber loadings but decreased with increase in fiber size. Water sorption characteristics of the composites were generally found to be poor. Similarly, the ability of fiber to retard flame was found to be unsatisfactory within the composite. Fiber loading for the experiments varied from 0% to 160% for all variable properties while fiber sizes were 0.150, 0.212 and 0.300 mm. Six percent weight of maleic anhydride addition showed maximum effect on the properties of composites.

Researching on mechanical behavior of OPMF and LLDPE composites, fiber size less than 300 µm and loading between 5 and 25 wt% were employed at 5 wt% interval [1]. Average tensile strength of pure LLDPE reduced by 36.78% as fiber loading increased from 5 to 25 wt%. Minimum tensile strength recorded was 7.27 MPa while the maximum was 8.97 MPa. Tensile modulus was also identified to enhance from 200 MPa of pure LLDPE at 0% fiber loading to 300 MPa at 25% loading. Thus, 25% increase in fiber loading enhanced tensile modulus by 50%. Nonetheless, percentage elongation at break reduced by 95.98% when loading was increased to 25 wt%. The respective change in elongation at break was found to be 87.83%, 89.17%, 94.82%, 95%, 76% and 95.98%. Examining the effect of loading on impact strength, the study identified reduction on values from 160.27 kJ/m^2 of pure LLDPE to 97.65 kJ/m^2 at 25% fiber loading, respectively, thus representing 16.58% (5 wt% fiber loading) to 39.07% (25 wt% fiber loading) reduction in impact strength. Comparatively, composite hardness decreased at 5 and 10 wt%, respectively. Nevertheless, at 20 and 25 wt%, there were respective increase by 18.63% and 24.56%. The respective hardness values were 14.96, 15.5, 20.60, 25.6 and 28.88 MPa at 5, 10, 15, 20 and 25 wt%, respectively. It was explained that the ability of fiber to deform interfacial bond due to higher stiffness quality and the presence of voids in composites as fibers become evenly dispersed contributed to the characteristic behavior of the OPMF–LLDPE composites.

4.12 Conclusion

Introducing petroleum-based plastics as composites with OPF in future research could be a greener way of dealing with polymer pollution and environmental management. OPF, particularly, bunch (OPEFB) fiber and OPMF have been found to mingle well with the manufacture of polymer composite materials. Those fibers are obtained from the oil palm tree, which has been identified to cope well in tropical regions such as Malaysia, Brazil, Nigeria and Ghana. With the current awareness to deal with environmental and issues associated with polymers, the favorable blend of OPFs and polymers in the formation of composites has been well embraced as a solution to environmental pollution. Bunch fiber is obtained after harvesting and removing fruits from the bunch. The mesocarp, which is directly attached to the fruit, is further processed before mesocarp fiber is obtained.

As composite material, OPF–polymer materials exhibit different properties depending on the type of fiber (bunch or mesocarp), size of fiber, percentage weight and/or volume of fiber and type of polymer material. Types of polymer materials include NR, PU, PS, EP, PES, PE and PVC as well as PP matrices. The presence of OPF in polymer matrices is mixed in terms of effects on composite properties. While some properties of composites are improved, others deteriorate. Similarly, with respect to the type of matrix, some properties either improved or deteriorated. The degree of improvement or deterioration also depends on the type of matrix. Therefore, there is more room for improvement in terms of OPF–polymer composite formation. The potential for improved technology, both material and processes, hence prevails.

In spite of these observations, studies on hybrid composites of OPF–polymer fabrications with other biomaterials are limited. The limitations are reflected, first, in terms of the type of second filler; second, type of polymer matrix; and third, the kind of composite properties that could be impacted. There is therefore immense potential with regard to composite formation technology and development as far as OPF–polymers are concerned. Polymer matrix consideration may also dwell on various thermosetting materials available in the waste stream. Finally, the effects of biofiber filler on variegated polymer matrix properties could be targeted during composite material research, for it is a truism that varying composite properties give credence to diversified material applications. Thus, for instance, studies relating to OPF–raw palm kernel–polymer composites, OPF–charred palm kernel–polymer composites or OPF–palm kernel ash–polymer composites may be considered in future studies. Studies also relating polymer composites involving PE, PP, PVC, PES, NR, EP or PU could be examined. Finally, hybrid composite properties with respect to physical, mechanical, thermal, electrical-electronic, chemical, environmental degradation/weathering, in terms of suitability of applications, might be investigated.

References

[1] Olusunmade, O.F., Adetan, D.A., and Ogunnigbo, C.O. A study on the Mechanical properties of oil palm mesocarp fiber-reinforced thermoplastics. J Compos 2016, 2016, 1–7. http://dx.doi.org/10.1155/2016/3137243.

[2] Motokeng, J.S. Comparison of injection moulded, natural fiber reinforced composites with PP and PLA as matrices, 2010, MSc thesis in Polymer Science. University of the Free State (Qwaqwa Campus).

[3] Shinoj, S., Visvanathan, R., Panigrahi, S., and Kuchubabu, M. Oil palm fiber (PF) and its composites: A review. Ind Crop Prod 2011, 33, 7–22.

[4] Hassan, S., Tesfamichael, A., and Nor, N.F.M. Comparison study of thermal insulation characteristics from oil palm fiber. MATECH Web of Conferences 2014, 13, 02016;. Doi: 10.1051/mateccouf/20141302016.

[5] Oladele, I.O., and Okoro, A.M. The effect of palm kernel shell ashon the mechanical properties as-cast aluminium alloy matrix composites. Leonardo Science Series 2016, 28, 15–30.

[6] Bledzk, A.K., Mumun, A.A., Jaszkiewicz, A., and Erdmann, K. Polypropylene composites with enzyme modified abaca fiber. Compos Sci Technol 2010, 70, 854–860. Doi: 10.1016/j.composcitech. 2010.02.003.

[7] Mishsra, S. et al. Mohanty, A.K., Drzal, L.T., Misra, M. and Hinrichsen, G., A review on pineapple Leaf fibers, sisal fibers and their biocomposites. Macromol Mater Eng 2004, 289, 955–974.

[8] Li, Y., Mai, Y-W., and Ye, L. Sisal fiber and its composites: A review of recent developments. Compo Sci Technol 2000, 60, 2037–2055.

[9] Bisanda, E.N.T., and Ansell, M.P. Properties of sisal-CNSL composites. J Mater Sci 1992, 27, 1690–1700.

[10] Silva, F.A., Chawla, N., and Filho, R.D. An experimental investigation of the fatigue behavior of sisal fibers. Mater Sci Eng A 2009, 516, 90–95.

[11] John, M.J., and Thomas, S. Biofibers and biocomposites. Carbohydr Polym 2008, 71, 343–364.

[12] Rozman, H.D., Saad, M.J., and Ishak, Z.A.M. Flexural and impact properties of oil palm empty fruit bunch (EFB) – Polypropylene composites-the effect of maleic anhydride chemical modification of EFB. Polym Test 2003, 22, 335–341.

[13] Hill, C.A.S.,&., and Khalil, H.P.S.H. b. Effect of fiber treatments on mechanical properties of coir or oil palm fiber reinforced polyester composites. J Apply Polym Sci 2000, 78, 1685–1697.

[14] Sreekala, M. S., Kumaran, M.G., Geethakumariamma, M.L., and Thomas, S. Environmental effects in oil palm fiber reinforced phenol formaldehyde composites: Studies on thermal, biological, moisture, and high energy radiation effects. Adv Compos Mater 2004, 13, 171–197.

[15] Raju, G., Ratman, C.T., Ibrahim, N.A., Rahman, M.Z.A.K., and Yunus, W.M.Z.W. Enhancement of PVC/ENR blend properties by poly (methyl acrylate) grafted oil palm empty fruit bunch fiber. J Appl Polym Sci 2008, 110, 368–375.

[16] Hariharan, A.B.A., and Khalil, H.P.S.A. Lignocellulose-based hybrid bilayer laminate composite: Part 1-studies on tensile and impact behavior of oil palm fiber-glass fiber-reinforced epoxy resin. J Compos Mater 2005, 39(8), 663–684.

[17] Yousif, B.F., and Tayeb, e.N.S.M. High-stress three-body abrasive wear of treated and untreated oil palm fiber-reinforced polyester composites. Proc Inst Mech Eng Part J J Eng Tribol 2008, 222, 637–646.

[18] Hill, C.A.S., Khalil, H.P.S.H., and Hale, M.D. A study of the potential of acetylation to improve the properties of plant fibers. Ind Crop Prod 1998, 8, 53–63.

[19] Feng, Y.H., Li, Y.J., Xu, B.P., Zhang, D.W., Qu, J.P., and He, H.Z. Effect of fiber morphology on rheological properties of plant fiber reinforced poly (Butylene succinate) composites. Compos B. 2013, 44, 193–199.

[20] Azwo, Z., Yousif, B., Manalo, A., and Karanasena, W. A review on the degradability of polymeric composites based on natural fibers. Mater Des 2013, 47, 424–442.

[21] Fu, S-Y., Yue, C-Y., Hu, X., and Mai, Y-W. Analysis of the micromechanics of stress transfer in single-and multi-fiber pull-out tests. Compos Sci Technol 2000, 60, 569–579.

[22] Chirajil, C.J., Joy, J., Mathew, L., Koetz, J., and Thomas, S. Nanofibril reinforced unsaturated polyester nanocomposites: Morphology, mechanical and barrier properties, viscoelastic behavior and polymer chain confinement. Ind Crop Prod 2014, 56, 246–254.

[23] Avancha, S., Beheral, A.K., Sen, R., and Adhikari, B. Physical and mechanical characterization of jute reinforced boy composites. J Reinf Plast Comopos 2013, 32(18), 1380–1390.

[24] Georgiopoulos, P., Kontou, E., and Christopoulos, A. Short-ter creep behavior of a biodegradable polymer reinforced with wood-fibers. Compos Part B: Eng 2015, 80, 134–144.

[25] Goriparthi, B.K., Suman, K., and Rao, N.M. Effect f fiber surface treatments on mechanical and abrasive near performance of polyactide/jute composites. Compos part A: Appl Sci Manuf 2012, 43(10), 1800–1808.

[26] Gurunathan, T., Mohanty, S., and Nayak, S.K. A review of the recent developments in biocomposites based on natural fibers and their application perspectives. Compos Part A: Appl Sci Manuf 2015, 77, 1–25.

[27] Hamma, A., Kaci, M., Ishak, Z.M., and Pegoretti, A. Starch-grafted-polypropylene/Kenaf fiber composites. Part 1: Mechanical performances and viscoelastic behavior. Compos Part A: Appl Sci Manuf 2014, 56, 328–335.

[28] Hao, A., Chen, Y., and Chen, J.Y. Creep and recovery behavior of Kenaf/polypropylene nonwoven composites. J Appl Polym Sci 2014, 111, 17. Doi: 10-1002/app.40726.

[29] Hidalgo-Salazar, M.A., Mina, J.H., and Herrera-Franco, P.J. The effect of interfacial adhesion on the creep behavior of LDPE-AL-Fique composite materials. Compos Part B: Eng 2013, 55, 345–351.

[30] Jimit, R.H., Zakaria, K.A., and Bapokutty, O. Influence of fiber orientation on mechanical properties of fiberglass reinforced composite. Centre for Advanced Research on Energy, Proceedings of Mechanical Engineering Research, 2017, 318–319.

[31] Kabir, M., Wang, H., Lau, K., and Cardona, F. Chemical treatments on plant-based natural fiber reinforced polymer composites: An overview. Compos Part B: eng 2012, 43(7), 2773–2892.

[32] Khan, J.A., Khann, M.A., and Islam, R. Mechanical, thermal and degradation properties of jute fabric-reinforced polypropylene composites: Effect of potassium permanganate as oxidizing agent. Polym Compos 2013, 34(5), 671–680.

[33] Kicinska-Jakubowska, A., Bagacz, E., and Zimniewska, M. Review of natural fibers Part I-vegetable fibers. J Nat Fibers 2012, 9(3), 150–167.

[34] Li, M., Li, D., Wnag, L.J., and Adhikari, B. Creep behavior of starch-based nanocomposite films with cellulose nanofibrils. Carbohydr Polym 2015, 117, 957963.

[35] Kalia, S., Thakur, K., Celli, A., Kiechel, M.A., and Schauer, C.L. Surface modification of plant fibers using environment friendly methods for their application in polymer composites, textile industry and antimicrobial activities. A review. J Environ Chem Eng 2013, 1(3), 97–112.

[36] Iu, K.C. The Application of oil palm fiber on concrete structure materials, 2015, BSc Mechanical Engineering (Designing and Innovation) Report, UniversitiTeknikal Malaysia Melaka.

[37] Chandramohan, D., and Marimuthu, K. Tensile and hardness tests on natural fiber reinforced polymer composite material. Int J Adv Eng Sci Technol 2011, 6(1), 097–104.

[38] Malasia Digest. 2014. Focus on oil palm under 11th Malaysia plan.www.malaysiandigest.com
[39] MPOB 2014. Oil palm planted area as at Dec 2013. bepi.mpob.gov.my
[40] Khalil, H.P.S.A., Siti, M.A., Ridzuan, R., Kamarudin, H., and Khairul, A. Chemical composition, morphological characteristics, and cell wall structure of Malaysian oil palm fibers. Polym Plast Technol Eng 2008b, 47, 273–280.
[41] Joseph, S., Joseph, K., and Thomas, S. Green composites from natural rubber and oil palm fiber: Physical and Mechanical properties. Int J Polym Mater 2006, 55, 925–945.
[42] Poku, K. Small scale palm oil processing in Africa, FAO Agricultural Services Bulletin 148, 2002, Food and Agriculture Organization of the United Nations, Rome.
[43] Hasamudin, W., and Soom, R.M. Road making using oil palm fiber, 2002, Malaysian Palm Oil Board Information series 171. Malaysian Palm Oil Board, Kuala Lumpur, Malaysia.
[44] Karina, M., Onggo, H., Abudullah, A.H.D.&., and Syampurwadi, A. Effect of oil palm empty fruity bunch fiber on the physical and mechanical properties of fiber glass reinforced polymer resin. J Biol Sci 2008, 8, 101–106.
[45] Sreekala, M.S., Kumaran, M.G., and Thomas, S. Oil palm fibers: Morphology, chemical composition, surface modification, and mechanical properties. J Appl Polym Sci 1997, 66, 821835.
[46] Singh, G., Manohan, S., and Kanopathy, K. 1982. Commercial scale bunched mulching of oil palms. In: Pushparajah, E., Chew, P.S. (Eds)., Proceedings of 1981 International Oil Palm Conference. Kuala Lumpur, Malaysia, pp. 367–277.
[47] White, N.M., and Ansell, M.P. Straw-reinforced polyester composites. J Mater Sci 1983, 18, 1549–1556.
[48] Wirjosentono, B.i., Guritno, P., and Ismail, H. Oil palm empty fruit bunch filled polypropylene composites. Int J Polym Mater 2004, 53, 295–306.
[49] Then, Y.Y. 2013 Ibrahim, N.A., Zainuddin, N., Ariffin, H. & Unus, W.M.Z.W. Oil palm mesocarp fiber as new lignocellulosic material or fabrication of polymer/fiber biocomposites. Int J Polym Sci 2013, 2013, Article ID 797452, 7.
[50] Jayashree, E., Manda, P.K., Madhava, M., Kamarag, A., and Sireesha, K. 2002. Development of decorticator for extraction of quality fiber from oil palm empty fruit bunches. In: Sreedharan, K., Kumar, P.K.V., Jayarama, Chulaki, B.M. (eds), Proceedings of the 15th Plantation Crops Sympos-ium-PLACROSYMXV. Chikmagalur, India, pp.10–13.
[51] Sreekala, M.S., Kumaran, M.G., and Thomas, S. Water sorption in oil palm fiber reinforced phenol formaldehyde composites. Compos Part A 2002b, 33, 763–777.
[52] Ibraham, N.A., Ilaiwa, F.A., Rahman, M.Z.A., Ahmad, M.B., and Dahlan, K.Z.M. Yunus, W.M.Z.W. Graft copolymerization of acrylamide onto oil palm empty fruit bunch (OPEFB) fiber. J Polym Res 2005, 12, 173–179.
[53] Yousif, B.F., and Tayeb, E.N.S.M. The effect of oil palm fibers as reinforcement on tribological performance of polyester composites. Surf Rev Lett 2007, 14(6), 1095–1102.
[54] Law, K.N., Daud, W.R.W., and Ghazali, A. Morphological and chemical nature of fiber strands of oil palm empty-fruit bunch (OPEFB). BioResources 2007, 2(3), 351–362.
[55] Chang, J.S.L., Chan, V.S., Law, M.C., and Leio, C.P. 2017 Comparative microstructure study of oil palm fruit bunch fiber, mesocarp and kernels after microwave pre-treatment. International Conference on Materials Technology and Energy. 217, 012025. Doi: 10.1088/1757-899X/217/012026.
[56] John, M.J., Francis, B., Varughese, K.T., and Thomas, S. Effect of chemical modification on properties of hybrid fiber biocomposites. Compos Part A 2008, 39, 352–363.
[57] Ramli, R., Haler, S., and Janialudin, M.A. Properties of medium density fiber board from oil palm empty fruit bunch fiber. Oil Palm Res 2002, 14(2), 34–40.

[58] Nordin, N.I.A.A., Ariffin, H., Andou, Y., Hassan, M.A., SHirai, Y., Nishida, H. et al. modification of oil palm mesocarp fiber characteristics using superheated steam treatment. Molecules 2013, 18, 9132–9146. Doi: 10.3390/molecules 18089132.
[59] Rozman, H.D., Tay, G.S., Kumar, R.N., Abusamah, A.I., Ismail, H., and Ishak, Z.A.M. The effect of oil extraction of the oil palm empty fruit bunch on the mechanical properties of polypropylene-oil palm empty fruity bunch-glass fiber hybrid composites. Polym Plast Technol Eng 2001b, 40, 103–115.
[60] Abubakar, A., Hassan, A., and Usof, A.F.M. The effect of oil extraction of the oil palm empty fruit bunch on the processability, impact, and flexural properties of PVC-U composites. Int J Polym Mater 2006, 55, 628–641.
[61] Rozman, H.D., Hilme, K.R.A., and Abubakar, A. Polyuretyane composites based on oil palm empty fruit bunches: Effect of isocyanate/hydroxyl ratio and chemical modification of empty fruit bunches with toluene diisocyanate and hexamethylenediisocyanate on mechanical properties. J Appl Polym Sci 2007, 106, 2290–2297.
[62] Khalil, H.P.S.A., Azura, M.N., Issam, A.M., Said, M.R., and Adawi, T.O.M. Oil palm Empty fruit bunches (OPEFB) reinforced in monounsaturated polyester composites. J Reing Plast Compos 2008a, 27(915 – 17), 1817–1826.
[63] KHOO, K C and LEE, T W (1991). Pulp and paper from the oil palm. Appita, 44(6): 385–388.
[64] Teuber, L., Osburg, V., Toprowski, W., Militz, H., and Krause, A. Wood polymer composites and their contribution to cascading utilization. J Clean Prod 2015, 110, 9–15.
[65] He, C., Hou, R., Xue, J., and Zhu, D. The performance of polypropylene wood plastic composites with different rice straw contents using two methods of formation. For Prod J 2013, 63, 61–66.
[66] McCormic, K., and Kautto, N. The bioeconomy in Europe: An overview. Sustainability 2013, 5, 2589–2608.
[67] US Energy Information Administration 2013. Annual Energy Outlook 2013 with Projections to 2040, DOE/E1A-0383(2013).
[68] Gonzàlez-Benito, J., and Gonzàlez-Benito, O. An analysis of the relationship between environmental motivations and ISO 14001 certification. Br J Manage 2005, 16, 133–148.
[69] Sreekala, M.S., Kumaran, M.G., Joseph, R., and Thomas, S. Stress-relaxation behavior in composites based on short oil-palm fibers and phenol formaldehyde resin. Compos Sci Tech 2001, 61, 1175–1188.
[70] Khalil, H.P.S.A., and Ismail, H. Effect of acetylation and coupling agent treatments upon biological degradation of plant fiber reinforced polyester composites. Polym Test 2001, 20, 65–75.
[71] Ewulonu, C.M., and Igwe, I.O. Properties of oil palm empty fruit bunch fiber filled high density polyethylene. Int J Eng Technol 2012, 3, 458–471.
[72] Yusoff, M.Z.M., and Salit, M.S. Mechanical properties of short random oil palm fiber reinforced epoxy composites. Sains Malaysiana 2010, 39(1), 87–92.
[73] Chollakup, R., Smitthipong, W., Kongtud, W., and Tantatherdtam, R. Polyethylene green composites reinforced with cellulose fibers (coir and palm fibers): Effect of fiber surface treatment and fiber content. J Adhes Sci Technol 2013, 27(12), 1290–1300. http://dx.doi.org/10.1080/01694243.2012.694265.
[74] Jacob, M., Jose, S., Thomals, S., and Varughese, K.T. Stress relaxation and thermal analysis of hybrid biofiber reinforced rubber biocomposites. J Reinf Plast Compos 2006b, 25, 5538–5547.
[75] Jacob, M., Varughese, K.T., and Thomas, S. Water sorption of hybrid bio-fiber-reinforced natural rubber biocomposites. Biomacromolecules 2005, 6(6), 2969–2969.

[76] Jacob, M., Varughese, K.T., and Thomas, S. Dielectric characteristics of sisal-oil palm hybrid biofiber reinforced natural rubber biocomposites. J Mater Sci 2006c, 41, 5538–5547.
[77] Mahjoub, R., Yatim, J.B.M., and Sam, A.R.M. A review of structural performance of oil palm empty fruit bunch fiber in polymer composites. Adv Mater Sci Eng 2013, Article ID 415359, 9. http:.//dx-doi.org/10.1155/2013/415359.
[78] Khalid, M., Ratman, C.T., Chuah, T.G., Ali, S., and Choong, T.S.Y. Comparative study of polypropylene composites reinforced with palm oil empty fruit bunch fiber and oil palm derived cellulose. Mater Des 2008, 29, 173–178.
[79] Rozman, H.D., Tay, G.S., Abubakar, A., and Kumar, R.N. Tensile properties of oil palm empty fruit bunch-polyurethane composites. Eur Polym J 2001a, 37, 1759–1765.
[80] Rozman, H.D., Tay, G.S., Abubakar, A., and Kumar, R.N. A preliminary study on the oil palm empty fruit bunch-polyurethane (EFB-PU) composites. Int J Polym Mater 2002, 51, 1087–1094.
[81] Badri, K.H., Othman, Z. B. and Razali, I.M. Mechanical properties of polyurethane composites from oil palm resources. Iran Polym J 2005, 14(5), 441–448.
[82] Bakar, A.A., and Baharulrazi, N. Mechanical properties of benzoylated oil palm empty fruit bunch short fiber reinforced poly(vinyl chloride) composites. Polym Plast Technol Eng 2008, 47, 1072–1079.
[83] Abubakar, A., Hassan, A., and Yusof, A.F.M. Effect of oil palm empty fruit bunch and acrylic impact modifier on mechanical properties and processability of unplasticized poly (vinyl) chloride composites. Polym Plast Technol Eng 2005, 44, 1135–1137.
[84] Chin, L.S. Characterization of natural fiber polymer composites for structural application. (dissertation), 2008, University Technology Malaysia, Johor Bahru, Malaysia.
[85] Lee, M.W., Han, S.O., and Seo, Y.B. Red algae fiber/poly (butylene succinate) biocomposites: The effect of fiber content on their mechanical and thermal properties. Compos Sci Technol 2008, 68, 1266–1272.
[86] Khalil, H.P.L.S.A., Hanida, S., Kang, C.W., and Fuaad, N.A.N. Agro-hybrid composite: The effects on mechanical and physical properties of oil palm fiber (EFB/glass hybrid reinforced polymer composites. J. Reinf. Plast. Compos. 2007, 26(2), 203–217.
[87] Zakaria, S., and Poh, L.K. Polyesterene-benzoylated EFB reinforced composites. Polym Plast Technol Eng 2002, 41, 951–962.
[88] Bakar, M.A.A., Natarajan, V.D., Kalaw, A., and Kudiran, N.H. 2007. Mechanical properties of oil palm fiber reinforced epoxy for building short span bridge. In: E.E. Gdoutos, (Ed.), Proceedings of the 13th Internatioanl Conference on Experimental Mechanics, July 1–6. Alexandroupolis, Greece, pp. 97–97.
[89] Hassan, A., Salema, AA., Ani, F.N., and Bakar, A.A. A review on oil palm empty fruit bunch fiber-reinforced polymer composite materials. Polym. Compos. 2010, 31(12), 2079–2101.
[90] Lalam, A., Sahari, B.B., Khalid, Y.A., and Wong, S.V. Fatigue behavior of oil palm fruit bunch fiber-epoxy and carbon fiber-epoxy composites. Compos Struct 2005, 71, 34–44.
[91] Shinoj, S., Visvanathan, S., and Panigrahi & Varadharaju, N. Dynamic mechanical properties of oil palm fiber (OPF)- linear low density polyethylene (LLDPE) biocomposites and study of fiber-matrix interactions. Biosystems Engineering 2011, 109, 99–107.
[92] Jacob, M., Francis, B., Thomas, S., and Varughese, K.T. Dynamical mechanical analysis of sisal/oil palm hybrid fiber-reinforced natural rubber composites. Polym Compos 2006a, 27, 671–680.
[93] Ghazilan, A.L.A., Mokhtar, H., Dawood, M.S.I.S., and Ali, J. S. M. Tensile mechanical property of oil palm empty fruit bunch fiber reinforced epoxy composites. International conference on Mechanical, Automotive and Aerospace Engineering 2016, 184(012046). Doi: 10.1088/1757-899X/184/1/01046.

[94] Sathiskumar, T.P., Navaneethakrishnan, P., Shankar, S., Rajasekar, R., and Rajini, N. Characterization of natural fiber. J Reinf Plast Compos 2014, 32(1a), 1457–1476.
[95] Bismarck, A., Mishra, S., and Lampke, T. Plant fibers as reinforcement for green composites, A.K. Mohanty, M. Misra, L.T. Drzal, Eds., Natural Fibers Biopolymers and Biocomposites, 2005, CRC Press., Taylor & Francis Group, USA, 35–112.
[96] Carus, M., Elder, A., Dammer, L., Korte, H., Scholz, L., Essel, R., and Breitmayer, E. Wood plastic composites (WPC) and natural fiber composites (NFC): European and Global Markets, 2012 and Future Trends, 2015, Nova Institute for Ecology and Innovation, 2015–15.
[97] Zampaloi, M., Pourboghrat, F., Yankovich, S.A., Rodgers, B.N., Moore, J., Drzah, L.T. et al. Kenaf natural fiber reinforced polypropylene composites: A discussion on manufacturing problems and solutions. Compos: Part A. 2007, 38, 1569–1580.
[98] Väisänen, T., Haapal, A., Lappalainen, R., and Tompo, L. Utilization of agricultural and forest industry waste and residues in natural fiber-polymer composites: A review. Waste Manage 2015, 54, 62–73.

Section III: **Simulation and Experimentation**

Tanya Buddi, Swadesh Kumar Singh and B. Nageswara Rao

5 Numerical simulations on the bio-based adhesive plywood house structure subjected to self-weight and wind loads

Abstract: Plywood, particleboards and medium-density fiberboards are currently used in construction of houses, furniture, partitions and others employing harmful formaldehyde as an adhesive material. Hazardous toxic flames emit out and people die when such plywood houses are subjected to fire. Bio-based plywood house constructions are environmentally feasible, potential, economically viable and healthy places to live and work. This chapter deals with a bio-based soy meal adhesive (SMA) for plywood manufacturing. A comparative study is made by evaluating the flexural strength of the commercially available formaldehyde-based glue-laminated plywood performing three-point bend test. The developed bio-adhesive has been characterized using field emission scanning electron microscopy and X-ray diffraction. The Taguchi's L_9 orthogonal array is adopted and performed ANOVA to trace the optimal plywood manufacturing parameters (viz. temperature and pressure) to achieve the maximum modulus of rupture. Finite element analysis has been carried out on the three-point bend test specimens using ANSYS and developed the stress–stain curves for soy meal composite and formaldehyde composite from the recorded load versus displacement data. STAAD-Pro software is utilized for analyzing a house structure construction considering self-weight and wind loads. SMA provides good flexural strength when compared to that of formaldehyde.

Keywords: Bio-adhesive, FESEM, formaldehyde, plywood, soy meal adhesive, XRD

5.1 Introduction

Natural resources are being exploited substantially to synthetic materials with increasing population and standard of living. High-cost carbon fiber-reinforced plastic composites are being used in sensitive aerospace applications. Low-cost, high-strength and lightweight wood and wood-based composites are being considered in other

Tanya Buddi, Department of Mechanical Engineering, Koneru Lakshmaiah Education Foundation, Deemed to be University, Green Fields, Guntur, India; Department of Mechanical Engineering, Gokaraju Rangaraju Institute of Engineering and Technology (GRIET), Hyderabad, India.
Swadesh Kumar Singh, Department of Mechanical Engineering, Gokaraju Rangaraju Institute of Engineering and Technology (GRIET), Hyderabad, India.
B. Nageswara Rao, Department of Mechanical Engineering, Koneru Lakshmaiah Education Foundation, Deemed to be University, Green Fields, Guntur, India

https://doi.org/10.1515/9783110655049-005

applications. Adhesives play a major role in wood composites. Formaldehyde, urea-formaldehyde and melamine-modified urea-formaldehyde resins are preferred for producing panels and building materials [1–3]. Exposure of these materials may enhance the risk of developing cancer [4, 5]. Japan and America have taken precautionary steps to minimize the emissions of formaldehyde from wood composites and introduced standards [6]. Environmental Protection Agency has banned the use of formaldehyde-based adhesives [7]. Advanced research is focused on minimizing the formaldehyde level from wood composites [8]. Bio-based formaldehyde-free adhesives are developed with appropriate chemistry on condensed tannins [9], lignin [10], dextrin [11], vegetable oils [12] and soy flour as well as soy protein [13]. Soy-based adhesives are found to be the leading glue for manufacturing plywood [14]. Pressure and temperature are the main manufacturing process variables. Taguchi method [15– 17] can be utilized for identifying the optimal manufacturing process parameters in achieving high modulus of rupture (MOR) through three-point bending tests on plywood samples. Conventional manual methods in design of complicated and high-rise structures are time-consuming due to cumbersome calculations. Akshay et al. [18] have analyzed G + 1 story residential building using STAAD-Pro (a powerful tool for structural analysis and design developed by Bentley systems, Inc.) and recommended green materials for eco-friendly, energy-efficient and self-sustainable buildings.

There is a need for finding an appropriate chemistry to make the bio-based adhesives alternate to the hazardous petroleum-based adhesives. The main objectives of this chapter are to develop environmentally friendly bio-adhesive for plywood manufacturing with soy meal and confirm its flexural strength with the commercial formaldehyde-based glue-laminated plywood. Optimal process parameters to achieve maximum MOR will be identified adopting Taguchi approach and performing ANOVA analysis. The developed bio-adhesive is characterized using FESEM (field emission scanning electron microscopy) and XRD (X-ray diffraction). Using ANSYS and specifying the measured load-displacement data of the three-point bending test specimen, stress–strain curve of the plywood with soy meal adhesive (SMA) is generated through a finite element modeling. Finally, STAAD.PRO software is used for analyzing a house structure with the developed bio-adhesive plywood composites.

5.2 Preparation of soy meal adhesive

Using appropriate proportions of constituents, details on preparation of SMA in ref. [8] are briefly highlighted as follows. Powdered soy meal is soaked in water. The mix is kept thick to break down dry powder lump (if any). Pine oil is mixed for 3 min and added 50% NaOH solution. About 10 g of sodium hydroxide pellets is added to 250 mL water. To modify soy protein structure, sufficient quantity of alkali is added in the second time addition of water and mixed for 2 min. Soy flour is treated with strong alkali for breaking internal hydrogen bond of coiled protein molecules. Soy meal-based glue formulation contains preservative chemicals like chlorinated phenol to provide mold resistance in high humidity condition. The SMA after mixing the constituents is shown in Figure 5.1.

Figure 5.1: Soy meal adhesive (SMA) after mixing the constituents.

Bio-adhesive prepared from soy meal powder is free from harmful carcinogen (formaldehyde). The waste soy meal seeds are freely available for low cost in India. Such adhesives are free from microbial attacks and enhance wood product life. The details on the manufacturing and testing of plywood and particle board using these bio-adhesives will be presented in the forthcoming section.

5.3 Manufacturing of plywood

Plywood making process consists of various operations like log selection, conditioning, cutting and debarking, peeling, veneer drying, clipping, composer, hot pressing, sawing and sanding [19, 20]. For manufacturing plywood, eucalyptus veneers (EV) are procured from Parijat Agro wood Industries Pvt. Ltd. Chevela, Hyderabad, India. The sorted sections are fed into a thermostatic oven to minimize shrink and moisture content (MC). EV are dried to 8% MC through veneer drier. They are cut to $160 \times 80 \times 1.6$ mm along and across grains, and stacked up veneers to odd number as in Figure 5.2. Laying-up and gluing of assembled veneer sections for a particular run of plywood are manually processed.

Figure 5.2: Seven stacked-up veneers.

These veneers are stacked in such a manner to have same longitudinal or lateral grain structure in alternate veneers by applying a thin layer of the adhesive between veneers. The plywood thickness depends on the number of veneer layers. A hot press is utilized to seal veneer into a solid piece of plywood. This task is carried out on the hot pressing test rig. The experimental setup consists of die, compression plate, heater, a power source (hydraulic power source) to apply force on compression plate, control unit and data recording system. When the sheets are loaded, the press squeezes them together under 1–1.6 MPa pressure at 110–135 °C. For the applied pressure, one can expect good contact between layers of veneer and improve the strength due to curing of the glue. After 7 min, the press is opened to unload the sheets. The time and temperature depend on the plywood design, thickness of veneer, wood species and machine.

The rough sheets are passed through a set of saws for trimming. The specimen plywood sheets are cut to the sizes as per the ASTM standards to measure the MOR, compressive strength, MC and density. A minimum of three samples are made for combinations of pressure and temperature.

5.4 Testing of plywood

Two types of specimens as per ASTM standards considered for testing are

Type-1: $t > 6$ mm, $L = 16t$ mm; $b = 50$ mm.
Type-2: $t < 6$ mm, $L = 16t$ mm; $b = 25$ mm.

To control weight, specimens are kept in a conditioning chamber at 27 ± 2 °C and 65 ± 5% relative humidity. Using digital calipers, width (b) and thickness (t) are measured to an accuracy of 0.01 mm at three locations and noted the average value.

Tests performed for evaluating the mechanical and physical properties are three-point bending flexural testing, compression testing, water absorption (WA)

testing, density testing and thickness swelling (TS) testing. Three-point bending test is performed on Universal Testing Machine of 25 kN load capacity functioning on servohydraulic motor using the BISS software.

MOR is evaluated from

$$\mathrm{MOR} = \frac{3PL}{2bt^2} \tag{5.1}$$

where P (N) is the load; L (mm) is the length of the panel; b (mm) is the width of the panel; and t (mm) is the thickness of the panel. Undeformed and deformed plywood sample is shown in Figure 5.3.

Figure 5.3: Undeformed and deformed plywood samples.

Table 5.1 gives the MOR for plywood with SMA at different process parameters (viz., temperature and pressure). Soy-based plywood is found to have high MOR when compared to that of petroleum-based PF resins mixed with strong alkali.

Table 5.1: Modulus of rupture, MOR (MPa), for plywood with SMA.

Temperature (°C)	Pressure (MPa)			
	1.0	1.4	1.5	1.6
110	65.90	66.38	19.32	19.08
135	67.09	70.66	21.67	19.96

Figure 5.4 shows the comparison of MOR with the applied pressure for the plywood using SMA at 110 and 135 °C. Gradual increase in MOR is noticed with increasing pressure from 1 to 1.4 MPa and a drastic fall in MOR is observed after 1.4 MPa pressure at 110 and 135 °C. This drastic behavior is due to improper bonding of adhesive with veneers at high pressures. Such a situation, the adhesive rushes out from sides of veneers, and makes improper bonding formation.

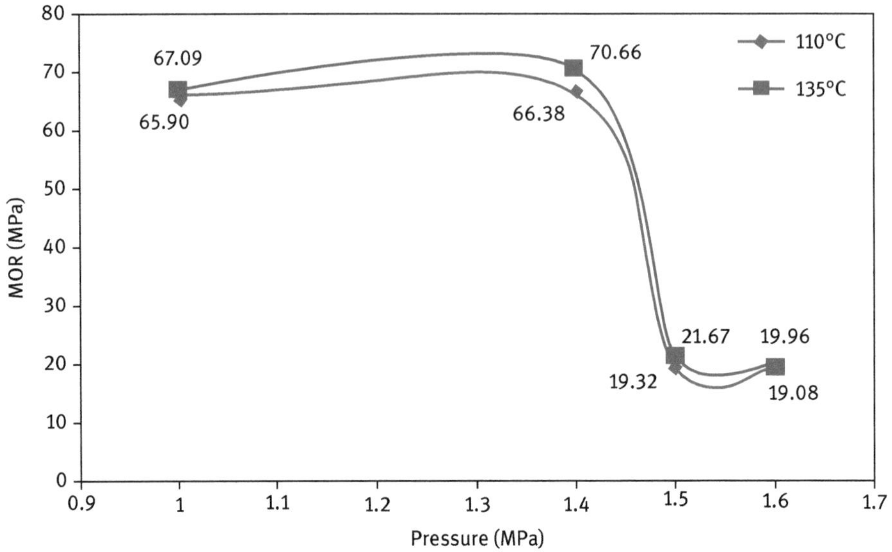

Figure 5.4: Comparison of MOR versus applied pressure for plywood with SMA.

It is recommended to make plywood with the developed bio-SMA processing under 1–1.6 MPa pressure at 110–135 °C. Optimal process parameters identified are 1.4 MPa pressure and 135 °C temperature.

Three specimens of 150 mm length and 50 mm width of the plywood and particleboard are tested. Compressive strength is evaluated from

$$\text{Compressive strength} = \frac{P}{bt} \tag{5.2}$$

where P (N) is the crushing load; b (mm) is the width of the specimen; and t (mm) is the thickness of the specimen.

For wood composite panels, TS and WA are primary measures of dimensional stability. The water soak test is done for 2 and 24 h to examine the behavior of the wood by the amount of water absorbed (WA) and the outcome of the consumed water on panel thickness by TS. These are measured by the change in weight and thickness of the panels before and after soaking in water. TS% is measured by soaking plywood or particleboard in water at room temperature for 2 h. For calculating dimensional stability, TS is worked out from

$$TS\% = 100 \times \frac{(t_w - t_{int})}{t_w} \tag{5.3}$$

where TS% is the TS percentage; t_w is the thickness of the wet sample at 2 h; and t_{int} is the initial thickness of the sample. Three sets of replicate panels are tested and an average value is considered. TS% values are measured for three types of bio-adhesives and are compared with those of commercial PF boards measured as per the IS: 12823-1990 standards.

WA is evaluated from

$$WA\% = 100 \times \frac{(W_w - W_{int})}{W_w} \tag{5.4}$$

where WA% is the WA percentage; W_w is weight of the wet sample at 24 h; and W_{int} is the initial weight of the sample. The WA of the average three samples is compared with those of PF boards measured as per the IS12823: 1990 standards.

The MC is evaluated for two specimens having 145 mm length and 50 mm width for each set of plywood. Using electronic balance (accuracy of 0.001 g), the weight of the specimen is measured as W_1. The specimen is placed in thermostatic air ventilated oven at 103 ± 2 °C for drying. The weight of the specimen is measured at regular intervals. The drying is complete when the variation in last two weights is <0.002 g. Denoting oven dry weight as W_0 g, the MC from

$$MC\% = 100 \times \frac{(W_1 - W_0)}{W_0} \tag{5.5}$$

The average value of three specimens is compared with those of PF boards measured as per IS12823: 1990 standards.

By using electrical balance, weight of the samples of each board is measured. The volume of the samples is calculated from the measured length, width and thickness using a digital caliper. Density (ρ) is determined from

$$\rho = \frac{m}{v} \tag{5.6}$$

where m is the mass and v is the volume of the wood panel. Table 5.2 gives comparison of properties of plywood with formaldehyde (PF) and plywood with SMA.

5.5 Microstructural observations

Characterization of the developed bio-adhesive has been characterized using FESEM and XRD utilizing a Zeiss MERLIN Compact. Sample surface is sputter coated with

Table 5.2: Comparison of plywood with different adhesive properties.

Property	Plywood with formaldehyde (PF)	Plywood with soy meal adhesive (SMA)
Modulus of rupture, MOR (MPa)	38.12	70.66
Compressive strength (MPa)	–	22.75
Thickness swelling, TS%	30.1	34.4
Water absorption, WA%	29.0	25.6
Density (kg/m^3)	755	617

20 nm thick gold layer. Observations are made with maximum acceleration voltage of 15 kV and highest spatial resolution of 3 nm. Micrograph of EV in Figure 5.5 clearly reveals micropores and the columnar form giving better provision for adhesive to enter into the veneers, thereby providing a fibrous type of structure with good bond strength. Figure 5.6 shows the micrographs of SMA with fine dispersive particles possesses higher MOR than the petroleum-based urea-formaldehyde [21].

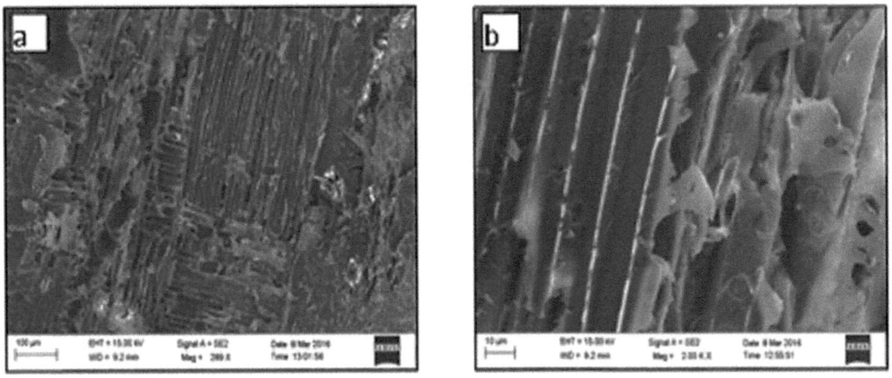

Figure 5.5: FESEM micrographs of EV.

Energy-dispersive spectroscopy (EDS) analysis has been carried out on EV and SMA [8]. It is noted that carbon presence may be due to hydrocarbons in wood and also using carbon tape. Oxygen (O) and gold (Au) found in EDS plots could be due to sputter gold coating and act as conducting medium in flow of electrons. EV consists of potassium along with ash, hemicellulose, cellulose and lignin. Dispersed particles in SMA matrix contain Ca, Mg and Si along with Al, Fe, Na and so on. Pd presence is due to soy matrix. Calcium is due to calcium sulfate or calcium carbonate presence in

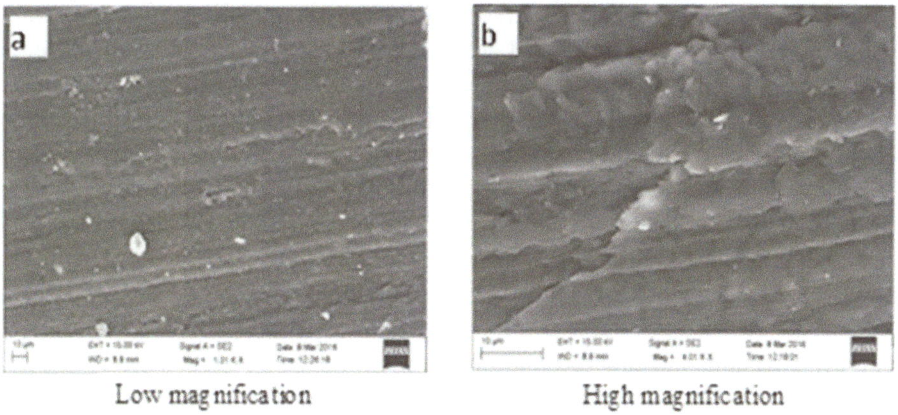

Figure 5.6: FESEM micrographs of SMA in the wood fibers.

preparatory layer [22]. Size and chemical composition variations in dispersed particles influence the adhesive properties.

X-ray powder diffraction analysis has been carried out on Bruker Discover D8 diffractometer using CuKα (λ = 1.5406 Å) graphite-monochromatized radiation. Patterns are obtained on oriented films by step scanning. They are from 5° to 40° 2θ (with a step size of 0.05° per 0.5 s), 40 kV and 40 mA in the X-ray tube. Figure 5.7 shows XRD profiles of adhesives, whose pattern confirms amorphous nature of adhesives.

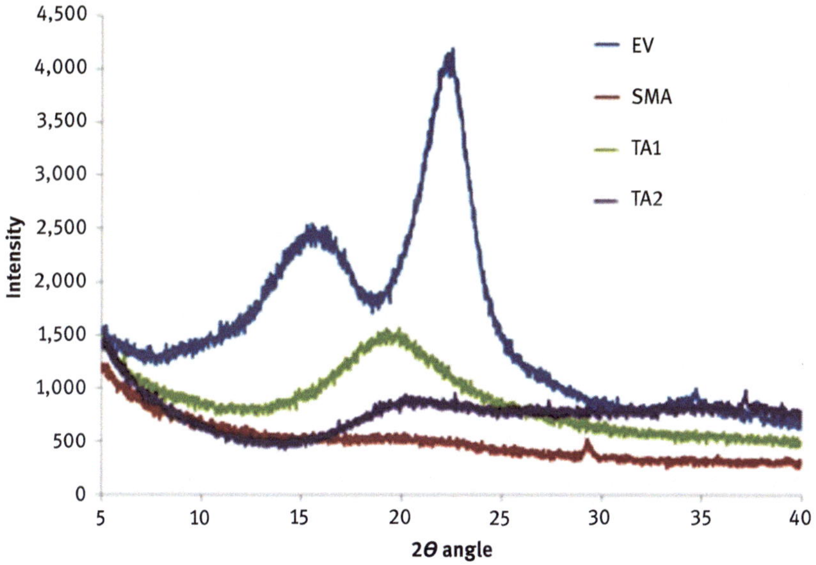

Figure 5.7: XRD patterns of EV and other adhesives (viz., soy meal adhesive (SMA), tamarind with formalin (TA_1) and tamarind with boric acid (TA_2)) with 2θ angle [8].

XRD pattern of EV in Figure 5.7 reveals peaks at 16° and 22°, whereas a different nature is observed in the developed adhesives. The amorphous nature of veneer is evident from broad peak widths. The amorphous nature of SMA with a small intensity peak is at 29.7°. The amorphous nature of TA_1 and TA_2 shows a similar trend up to 20° and differ afterward. TA_1 pattern is comparable with urea-formaldehyde pattern due to formalin use in preparation. The high intensity peak at 20° for TA_1 is wide when compared to the small intensity peaks of TA_2 and SMA. TA_2 pattern shows a small intensity peak at 38°. No peaks in similarity with formaldehyde indicate the absence of formaldehyde in TA_2 and SMA.

FESEM investigations reveal the presence of dispersed particles. XRD diffraction investigations indicate the amorphous nature of the developed bio-adhesives. Due to homogenous dispersed particles and matrix uniformity, SMA possesses good rupture strength when compared to that of formaldehyde.

5.6 Optimal plywood process parameters

Optimal plywood manufacturing process parameters (viz., pressure and temperature) to achieve maximum MOR in three-point bending tests are identified for the developed SMA utilizing a simple and reliable Taguchi L_9 orthogonal array [23]. In Taguchi method, number of experiments (N_{Taguchi}) for plywood process parameters (factors) and levels can be found from [24]

$$N_{\text{Taguchi}} = 1 + \text{Number of factors} \times (\text{Number of Levels} - 1) \tag{5.7}$$

Table 5.3 provides the levels of input process parameters such as pressure and temperature designated by A and B, respectively. MOR to the levels of the process parameters in Table 5.4 are as per the Taguchi's L_9 orthogonal array. It should be noted that eq. (5.7) gives four factors for $N_{\text{Taguchi}} = 9$ (nine test runs) and three levels. As in ref. [25], two fictitious factors C and D are introduced to the two plywood manufacturing process parameters A and B. ANOVA is performed to trace the optimal manufacturing process parameters of the plywood for the maximum MOR.

Table 5.3: Assigned levels of the process parameters in plywood manufacturing using SMA.

Control factors	Designated factor	Level 1	Level 2	Level 3
Pressure (MPa)	A	1.0	1.2	1.4
Temperature (°C)	B	110	120	135
Fictitious	C	c_1	c_2	c_3
Fictitious	D	d_1	d_2	d_3

Table 5.4: Modulus of rupture (MOR) for test runs as per the Taguchi's L_9 orthogonal array.

Test run	Levels of input parameters				Modulus of rupture, MOR (MPa)			
					Test	Analysis – eq. (5.8)		
	A	*B*	*C*	*D*		Only process parameters	Inclusion of fictitious parameters	Expected range
1	1	1	1	1	65.90	65.18	65.90	64.40–66.00
2	1	2	2	2	66.35	66.29	66.35	65.51–67.11
3	1	3	3	3	67.09	67.87	67.09	67.09–68.69
4	2	1	2	3	66.24	66.18	66.23	65.40–67.00
5	2	2	3	1	67.28	67.29	67.28	66.51–68.11
6	2	3	1	2	68.84	68.87	68.83	68.09–69.69
7	3	1	3	2	66.38	67.14	66.37	66.36–67.96
8	3	2	1	3	68.20	68.25	68.20	67.47–69.07
9	3	3	2	1	70.66	69.83	70.65	69.05–70.65

ANOVA results in Table 5.5 give the contribution of the plywood manufacturing process parameters on the grand mean value of MOR as 30.1% and 57.3%, respectively. To get maximum MOR, the optimal manufacturing process parameters identified from the ANOVA Table 5.5 are A_3B_3 in which subscripts to the parameters A and B denote the level. Conformation experiments are mandatory. The additive law [24] is followed to estimate MOR.

Table 5.5: ANOVA for MOR (MPa) of the plywood using soy meal adhesive.

Parameters	1-Mean	2-Mean	3-Mean	Sum of squares	% contribution
A	66.45	67.45	68.41	5.763	30.1
B	66.17	67.28	68.86	10.975	57.3
C	67.65	67.75	66.92	1.232	6.4
D	67.95	67.19	67.18	1.171	6.1

In case of two plywood manufacturing process parameters (A and B), the MOR estimated from the additive law is

$$\text{MOR} = (\bar{\text{M}}\bar{\text{O}}\bar{\text{R}})_{Al} + (\bar{\text{M}}\bar{\text{O}}\bar{\text{R}})_{Bl} - \text{MOR}_{\text{mean}} \tag{5.8a}$$

In the case of introducing two fictitious parameters (*C* and *D*) to the two plywood manufacturing process parameters (*A* and *B*), the MOR using the mean values of four parameters (*A*, *B*, *C* and *D*) in the ANOVA Table 5.5 from the additive law is

$$\text{MOR} = (\bar{\text{M}}\bar{\text{O}}\bar{\text{R}})_{Al} + (\bar{\text{M}}\bar{\text{O}}\bar{\text{R}})_{Bl} + (\bar{\text{M}}\bar{\text{O}}\bar{\text{R}})_{Cl} + (\bar{\text{M}}\bar{\text{O}}\bar{\text{R}})_{Dl} - 3\,\text{MOR}_{\text{mean}} \tag{5.8b}$$

Subscripts in eqs. (5.2a) and (5.2b) *A*1, *B*1, *C*1 and *D*1 refer to the mean values of *MOR* for the level of parameters. MOR_{mean} is the grand mean value of MOR.

For the identified optimal plywood manufacturing process parameters (A_3B_3), the MOR estimated from eq. (5.8a) is 69.83 MPa, whereas the ninth test result in Table 5.4 is 70.66 MPa. With inclusion of fictitious parameters, MOR is estimated from eq. (5.8b) as 70.65 MPa. Estimate of MOR for the identified optimal process parameters (A_3B_3) using eq. (5.8b) is found to be very close to the ninth test run result in Table 5.4. Considering the low and high mean values of MOR corresponding to the fictitious parameters (*C* and *D*) from Table 5.5, one can find the expected range of MOR. Table 5.4 presents the expected range of MOR for each test run. The test results are found to be within the expected range of MOR, which validates the estimates of the output response using the additive law.

From the mean values of MOR in Table 5.5 for the levels of manufacturing process parameters (*A* and *B*), an empirical relation is developed for MOR in the form

$$\text{MOR} = 44.46 + 0.61A + 0.163B - 0.005A^2 - 0.000227B^2 \tag{5.9}$$

It should be noted that the process parameter, pressure designated by *A*, is varying from 1.0 to 1.4 MPa, whereas another process parameter, temperature designated by *B*, is varying from 110 to 135 °C. The empirical relation (5.9) provides the MOR value for optimal process parameters of 135 °C temperature and 1.4 MPa pressure as 69.9 MPa. The expected range of MOR for the above optimal process parameters is from 69.05 to 70.65 MPa.

Empirical relation (5.9) can be utilized for estimation of MOR for the specific values of *A* and *B*. It is noted that estimates of MOR using the additive law (5.8) as well as the empirical relation (5.9) are reasonably in good agreement with experimental results. It is observed that MOR increases with increasing the pressure (from 1.0 to 1.4 MPa) and also with increasing temperature (from 110 to 135 °C).

Drastic reduction in MOR is noticed in test results of Table 5.6 for 1.5 and 1.6 MPa pressure. Hence, the above identified optimal manufacturing process parameters provide maximum MOR for the plywood using SMA.

Table 5.6: Modulus of rupture (MOR) of the plywood using soy meal adhesive for the specific process parameters.

Pressure, *A* (MPa)	Temperature, *B* (°C)	Modulus of rupture, MOR (MPa)
1.5	110	19.32
1.5	135	21.7
1.6	110	19.1
1.6	135	20.0

5.7 Generation of stress–strain curve

Finite element analysis (FEA) of the three-point bend specimen is performed utilizing ANSYS software package for the specified loads. The model is validated through comparison of FEA deflections with measured data for the applied load. The FEA deflections are in good agreement with test data. This confirms the modeling of the structure and specifications of the conditions (boundary conditions and loads). Figures 5.8 and 5.9 show the stress and strain contour plots. The stress versus strain curves are generated for the plywood with SMA (or soy meal composite, SMC) and plotted the stress versus strain curve from simulations in Figure 5.10. The maximum stress at a load of 1.1 kN is 66.2 MPa for SMC or plywood, whereas that of formaldehyde composite (FC) is 32.6 MPa. Using the stress–strain data of the plywood, it is possible to estimate the load bearing capacity of any complex plywood construction.

5.8 Modeling of a plywood house structure

Finite element modeling and analysis of a plywood house structure with SMA has been carried out using the commercial finite element software STAAD-Pro. Assessment is made on structural deformation (displacements), stresses, shear force, bending moment of a house structure subjected to self-weight and wind loads. To examine the adequacy of the STAAD-Pro software package, a simply supported beam is modeled and validated with design formula. Also, a three-point bending plywood test specimen is modeled and validated with test results.

A plywood house structure is modeled in STAAD-Pro software. The step-by-step procedure for generation of the structure with the required dimensions, specification of material properties and execution is explained with images captured from STAAD-Pro. FEA has been carried out for SMC and FC house structures having same dimensions.

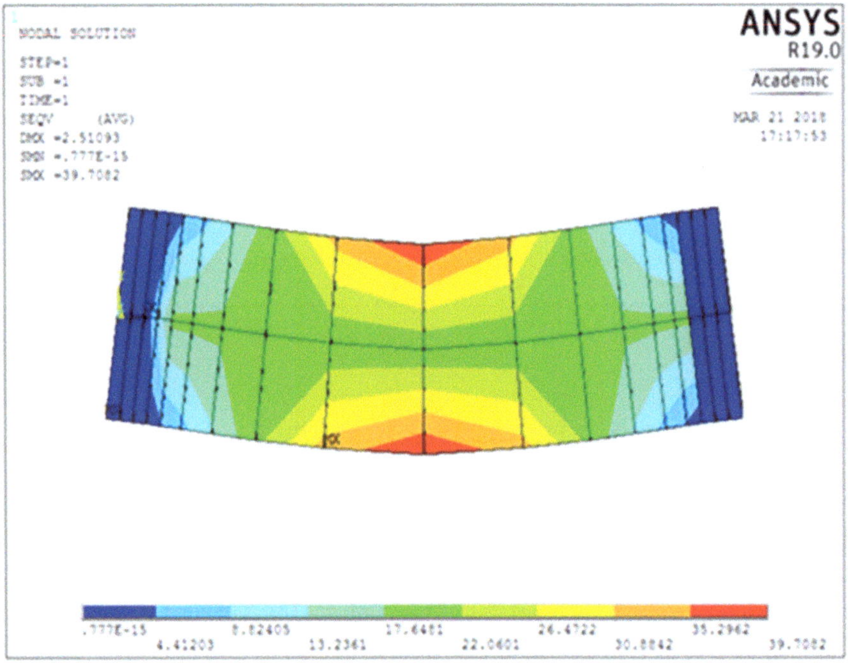

Figure 5.8: Stress contour plot at a load $P = 0.8$ kN and $E = 1{,}536$ MPa.

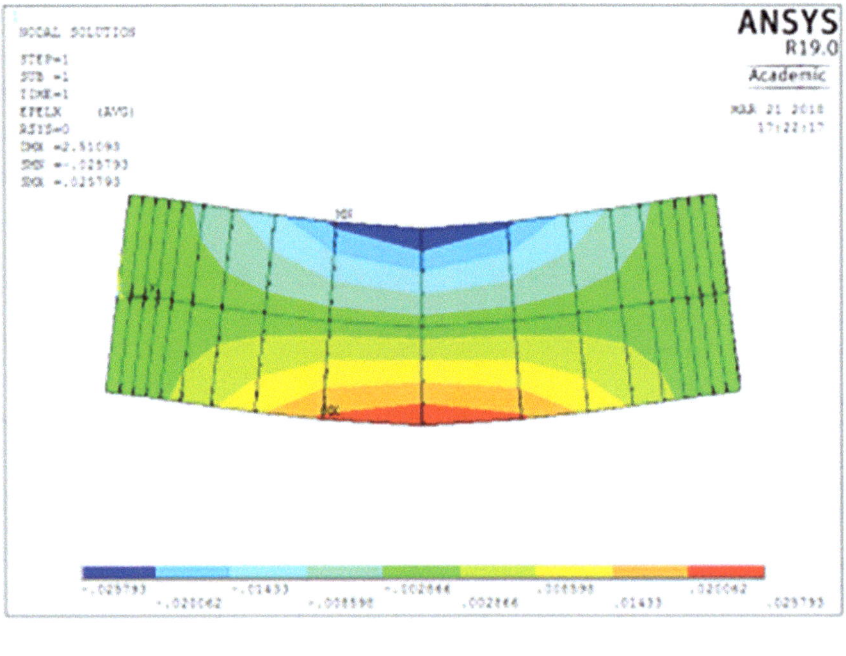

Figure 5.9: Strain contour plot at a load $P = 0.8$ kN and $E = 1{,}536$ MPa.

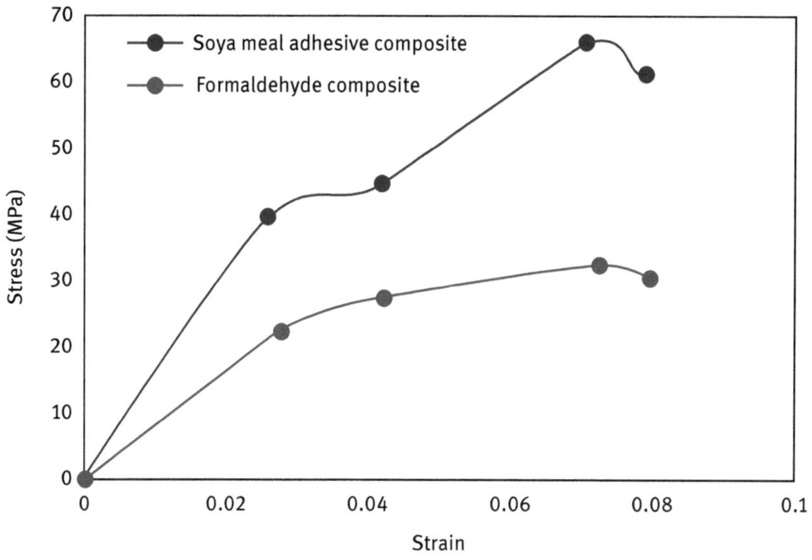

Figure 5.10: Stress versus strain curves for SMC and FC.

The house structure is generated in STAAD-Pro using structure wizard. A bay frame is selected in the frame model. Three bays are selected along length, width and height directions. The size of the structure is 3 × 3 × 3 m. In the next step, dimensions of the columns and beams are assigned as 0.45 × 0.23 m, 0.25 × 0.2 m, respectively. Internal columns and beams are constructed to support the walls. The thickness of slabs and walls is assigned as 0.15 and 0.23 m, respectively. Surface of the wall is generated. Similarly, all slabs and walls are generated for the house structure. Table 5.7 gives the specified material properties for SMC and FC.

Table 5.7: Material properties of SMC and FC.

Property	Soy meal composite (SMC)	Formaldehyde composite (FC)
Young's modulus (MPa)	1,536.3	1,600
Poisson's ratio	0.15	0.15
Density (kN/m^3)	5.591	7.405
Coefficient of thermal expansion	5.6×10^{-6}	5.6×10^{-6}
Critical damping	0	0
Shear modulus (MPa)	667.94	695.65
Bulk modulus (MPa)	731.55	761.90

After specification of material properties, the bottom surface of the house structure is fixed. The loads are generated by the STAAD-Pro load generator. Loading cases are categorized as self-weight and wind load. Wind loads are generated by software considering wind intensities at different heights as per the specifications of ASCE-7 code. Exposure factor of 0.8 is considered. The wind loads are evaluated for the building as a whole (individual structural elements as roofs and walls; and individual cladding units including glazing and their fixings). Wind intensities at heights are fed manually (to the software) for wind load generation at different floors.

The structure is analyzed for combinations of loads (self-weight and wind load). Job information indicates number of nodes = 577; highest node = 64; number of elements = 120; highest beam = 120; number of plates = 500; and highest plate = 620.

Shear force diagram (SFD) and bending moment diagram (BMD) under self-weight are shown in Figure.5.11, whereas Figure 5.12 shows the combined self-weight and wind load conditions.

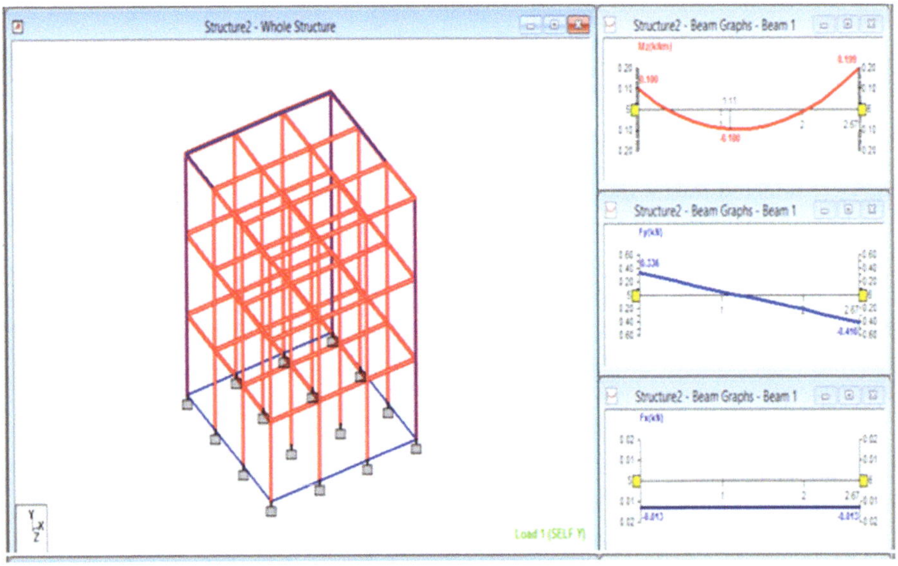

Figure 5.11: SFD and BMD under self-weight.

Similarly, FEA has been performed for FC. Table 5.8 shows that displacement of formaldehyde laminate and SMC is close. Reaction force of SMC is less when compared to that of FC due to more glue strength of SMA. Compressive and tensile strength of FC is slightly more than SMC. The bending moment and shear force of SMC in Table 5.9 is lesser than the FC laminate because of more flexural strength of SMC. Formaldehyde can be replaced with bio-adhesive SMA (free from carcinogenic emissions) because there is no difference in the structural response.

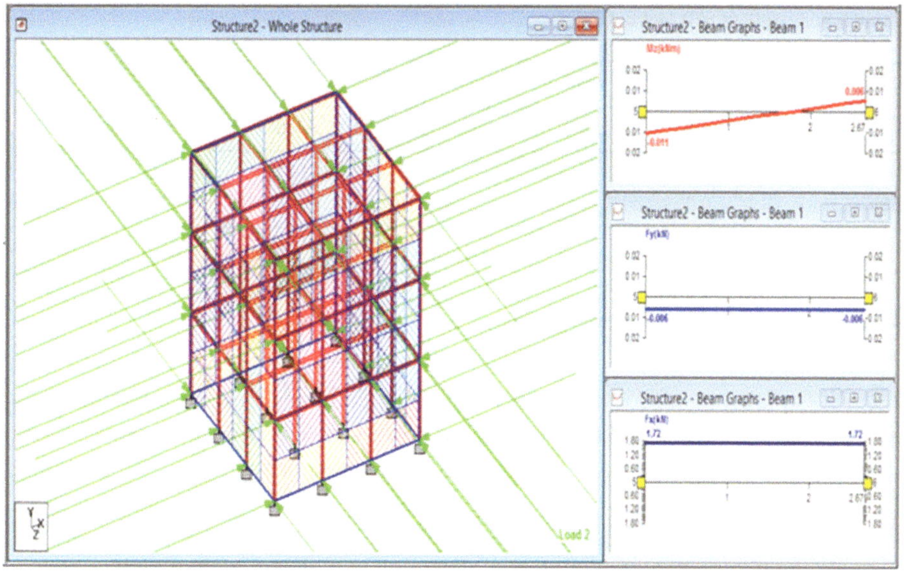

Figure 5.12: SFD and BMD under combined loading condition.

Table 5.8: Finite element results of SMC and FC house structures.

STADD-Pro results	FC	SMC
Maximum displacement (mm)	0.185	0.188
Reaction force along *X* (kN)	35.453	35.408
Reaction force along *Y* (kN)	154.548	153.021
Reaction force along *Z* (kN)	35.449	35.406
Moment along *X* (kN m)	3.952	3.955
Moment along *Y* (kN m)	0.001	0.001
Moment along *Z* (kN m)	3.948	3.953
Compressive stress (MPa)	291	286
Tensile stress (MPa)	178	171
Resultant surface force (kN)	64.647	64.583
Resultant surface moment (kN m)	181.583	181.102

Table 5.9: Results of SFD and BMD.

Composite	Bending moment (kN m) at			Shear force (kN) at		
	$x = 0$	$x = 1.1$	$x = 2.67$	$x = 0$	$x = 1.1$	$x = 2.67$
FC	0.133	−0.132	0.264	0.543	0	0.445
SMC	0.1	−0.1	0.199	0.41	0	0.336

5.9 Concluding remarks

Adhesives play a major role in wood composites. Formaldehyde, urea-formaldehyde and melamine-modified urea-formaldehyde resins are preferred adhesives for producing panels of exterior grade. Under elevated temperatures and high humid conditions, these formaldehyde emissions are suspected to cause throat and nasal congestions, burning eyes or headaches and even affect the upper respiratory system. Exposure of these materials may enhance the risk of developing cancer. This chapter is devoted to find an appropriate chemistry to realize the bio-based adhesives for replacing the traditional petroleum-based adhesives. In this regard, summary of findings made in the chapter are highlighted below.

- Bio-adhesives from the soy meal powder are free from a harmful carcinogen (formaldehyde), which are free from microbial attacks and expected to enhance the life of wood products. The waste soy meal and tamarind seeds are freely available or at low cost in India.
- The SMA possesses high MOR when compared to that of the petroleum-based urea-formaldehyde adhesive.
- Taguchi's L_9 orthogonal array is adopted and performed experiments to identify the optimal plywood manufacturing process parameters for achieving maximum MOR. The maximum MOR of 70.7 MPa is achieved for the processing conditions: 135 °C temperature and 1.4 MPa pressure.
- FESEM images reveal better rupture strength of soy adhesive when compared to formaldehyde due to matrix uniformity and homogenous dispersed particles.
- The XRD analysis on SMA reveals an amorphous nature with no wider peaks indicating absence of dangerous formaldehyde. Adhesives prepared from soy provide show better performance.
- From the stress–strain curves, it is observed that SMC possesses more strength than FC.
- Bending strength and shear strength of SMC are high when compared to that of FC.

- By using STAAD-pro software, compressive and tensile strengths of FC are comparable to soy due to the alignment of the layers in perpendicular direction and proved that this material is worthy.

Future research is directed toward the development of bio-based alternatives to timber for building and construction industries by improving microbial-resistant and fire-retarding properties utilizing nontoxic additives and treatments.

References

[1] Kim, S. "Environment-friendly adhesive for surface bonding of wood-based flooring using natural tannin to reduce formaldehyde and TOVC emission." Bio-resource Technol 2009, 100, 744–748.

[2] Halvarsson, S., Edlund, H., Norgren, M. "Properties of the medium density fibreboard (MDF) based on wheat straw and melamine modified urea-formaldehyde (UMF) resin." Indus Crop Prod (2008), 28 (1), 37–46.

[3] Mansouri, H.R., Pizzi, A. "Recycled micronized polyurethane powders as active extenders of UF and PF wood panel adhesives." HolzRoh- Werkst 2007, 65(4), 293–299.

[4] International Agency for Research on Cancer, 2004. Press Release No.153 – IARC Classifies Formaldehyde as Carcinogenic to Humans, retrieved on 28th January 2014. (http://www.iarc.fr/en/mediacentre/pr/2004/pr153.html).

[5] Sung, M., Lee, S.M. and Min Y. "Decreasing the formaldehyde concentration in indoor air by improving the adhesives used in engineered wood materials in Korean apartment buildings". J Adhe Sci Technol 2013, 27 (5–6), 671–682.

[6] Chengsheng, Gui., Jin, Zhu., Zhongtao, Zhang., and Xiaoqing, Liu. "Research progress on formaldehyde free wood adhesive derived from soy flour" Adhesives – Applications and Properties". http://dx.doi.org/10.5772/65502, (2016).

[7] EPA United States Environmental Protection Agency. EPA Issues Final Rule to Protect the Public from Exposure to Formaldehyde, (2017). (https://www.epa.gov/newsreleases/epa-issues-final-rule-protect-public-exposure-formaldehyde).

[8] Tanya Buddi, K. Mahesh, Nitin Muttil, Nageswara Rao B., Nagalakshmi J., Swadesh Kumar Singh, "Characterization of plywood produced by various bio-adhesives.", Elsevier Mater Today: Proc 2017, 4, 496–508.

[9] Pichelin, F., Nakatani, M., Pizzi, A., Wieland, S., Despres, A., and Rigolet S. "Structural beams from thick wood panels bonded industrially with formaldehyde-free tannin adhesives." Forest Prod J 2006, 56(5), 31–36.

[10] Ghaffar, S.H., and Fan M. "Lignin in straw and its applications as an adhesive." Int J Adhes Adhes 2014, 48, 92–101.

[11] Liu, X., Wang, Y., Cao, Y., Vikram, Yadama., Xian, M., and Zhang, J. "Study of dextrin-derived curing agent for waterborne epoxy adhesive". Carbohydr Polym 2011, 83(3), 1180–1184. DOI: 10.1016/j.carbpol.2010.09.019.

[12] Pizzi A., "Recent developments in eco-efficient bio-based adhesives for wood bonding: opportunities and issues." J Adhes Sci Technol (2006), 20, 829–846.

[13] Mo, X., and Sun, X.S. "Soy proteins as plywood adhesives: formulation and characterization". J Adhes Sci Technol 2013, 27(18–19), 2014–2026.

[14] Lépine, E., Riedl, B., Wang, X.M., Pizzi, A., Delmotte, L., Hardy, J.M., and Cruz, M.J.R.D. "Synthesis of bio-adhesives from soybean flour and furfural: Relationship between furfural level and sodium hydroxide concentration". Int J Adhes Adhes 2015, 55(5), 6270–6273. DOI: 10.1016/j.ijadhadh.2015.08.007.
[15] Ross, P.J. "Taguchi Techniques for Quality Engineering", McGraw-Hill, Singapore (1989).
[16] Srinivasa, Rao B., Rudramoorthy, P., Srinivas, S., and Nageswara Rao, B. "Effect of drilling-induced damage on notched tensile strength and pin-bearing strength of woven GFR-epoxy composites". Mater Sci Eng A 2008, 472, 347–352.
[17] Singaravelu, J. , Jeyakumar, D., and Nageswara Rao, B., "Taguchi's approach for reliability and safety assessments in the stage separation process of a multistage launch vehicle". Reliab Eng Syst Saf 2009, 94(10), 1526–1541.
[18] Akshay, B.M., Allaudin, I.S., Shamashree, S.R., Sushma, J.P., and Uday, J.P. "Green building materials– A way towards sustainable construction". Int J Appl or Innovat & Manage (IJAIEM) 2015, 4 (4), 244–249.
[19] Blackwell Duncan S., "The Complete Plywood Handbook", 1st Edition, Tab Books, Amazon (1981).
[20] Lutz, J.F. "Techniques for peeling, slicing and drying veneer", USDA Forest Service, Research Paper FPL 228, Forest Product Laboratory, Madison, Wisconsin (1974). https://www.fpl.fs.fed.us/documnts/fplrp/fplrp228.pdf.
[21] Carvalho, Ana P., Vaz, Maria F., Marta, M., Ferreira and João Pires. "Gilded wood from the organ of the Church of Santa Cruz (Coimbra)". J Brazil Chem Soc 2008, 19(8) . http://dx.doi.org/10.1590/S0103-50532008000800029.
[22] Ross, P.J. "Taguchi Techniques for Quality Engineering", McGraw-Hill, Singapore (1989).
[23] Tanya, Buddi., Swadesh Kumar, Singh., and Nageswara Rao, B. "Optimum process XE "parameters" for plywood manufacturing using soya meal adhesive". Elsevier Materials Today: Proceedings 2018, 5, 18739–18744.
[24] Srinivasa Rao, B., Rudramoorthy, P., Srinivas, S., and Nageswara Rao, B. "Effect of drilling induced damage on notched tensile strength and pin-bearing strength of woven GFR-epoxy composites". Mater Sci Eng A 2008, 472, 347–352.
[25] Rajeev Kumar, D., Varma, P.S.S.K., and Nageswara Rao, B. "Optimum drilling parameters of coir fiber-reinforced polyester composites". Am J Mech Ind Eng 2017, 2(2), 92–97.

Raghu Raja Pandiyan Kuppusamy

6 Isothermal mold filling simulations for developing liquid composite molded parts

Abstract: Polymer matrix composite materials are formed by combining two or more heterogeneous materials, namely polymer matrix and reinforcement. It's difficult to manufacture large, complex, three-dimensional structures using traditional manufacturing processes. Resin transfer molding (RTM) is introduced to meet the needs of the process and reduce complexity. Several models have been proposed that can help in either designing the mold or predicting the filling time, which are essential factors in the process. All models proposed use Darcy's law to simulate the resin flow inside the mold. Hence the key parameters of the RTM process are the process variables used in Darcy's law. They are inlet/outlet pressures, permeability, viscosity, mold shape and inflow rate or velocity of the resin. Different models have different approaches to estimate the filling rate or the flow-front position of the injected resin from these parameters. The main objective is to study and evaluate the various commercially available finite element method (software packages like ANSYS. The second is to develop an effective injection strategy, which includes location of vents and gates, max injection pressure so as to reduce the filling time and avoid formation of dry spots by performing mold filling simulations for industrial parts. The initial part of this work aims at evaluating the range of applicability of ANSYS to perform mold filling simulations. Specialized RTM packages like PAM-RTM will also be evaluated to overcome the shortcomings of the standard available packages. Mold filling simulations will be initially performed on simple objects. The latter part of the work aims at using the appropriate package to perform mold filling simulations and predict the effective injection strategy for industrially important objects that are currently being manufactured by hand layup.

Keywords: Mold filling simulations, liquid composite molding, resin transfer molding, Darcy's law, porous flow, injection strategy

6.1 Introduction

Polymer matrix composite materials are formed by combining two or more heterogeneous materials, namely polymer resin and reinforcement fibers. Polymer composites have been in use for decades. Their advantages over other materials for high-performance, lightweight applications have attracted many industries such

Raghu Raja Pandiyan Kuppusamy, Department of Chemical Engineering, National Institute of Technology Warangal, Telangana, India

https://doi.org/10.1515/9783110655049-006

as aerospace, automobile, infrastructure and marine. The path to the design and manufacturing of composite structures was pursued in evolutionary as well as revolutionary ways. They ranged from using hand layup (HLU) with labor and cost-intensive autoclave processing.

Among many manufacturing methods, HLU is most common and versatile. HLU involves manufacturing by the sequential addition of layers of reinforcement and resin matrix in an open mold manually. It allows for the manufacture of a wide range of geometries that require low initial investment. Despite these advantages, there are several limitations of the HLU process, which include low reinforcement volume fraction, nonuniform quality leading to uneven thickness and nonuniform distribution of reinforcement material and matrix. Longer production time, lower production rate and high involvement of skilled labor make the process unsuitable for large-scale production. Also, being an open mold process, it emits large volumes of styrene, which make the process environment unfriendly. Many new manufacturing techniques were invented and introduced during the last two decades. Manufacturing composite materials encounters many drawbacks. It's difficult to manufacture large, complex, three-dimensional (3D) structures using traditional processes. Resin transfer molding (RTM) is then introduced to meet the need of a process that can deal with structures with complicity. The process engineer has relied on experience and trial-and-error approaches to improve the manufacturability of a prototype. This has proved to be very expensive. Engineers have now resorted to use process modeling and simulations to address some of the concerns.

All liquid molding processes involve impregnation of the resin into a fibrous network bed. The goal in this process is to saturate all the empty spaces between the fibers before the resin gels. RTM consists of a mold cavity that is in the shape of the part to be manufactured. The fiber preform is placed in the cavity. The mold is closed and clamped and held under pressure with a press. The resin is then injected into the compressed preform through one or more gates from a pressurized container. Once full, injection is discontinued and the resin allowed curing. This cure may be initiated by either heating or use of inhibitors. The mold is opened once the cure is complete [1, 2]

Basically the wetting of reinforcement in an RTM process is a typical porous media flow problem. Interaction between liquid phase (resin) and porous medium (fiber preform) can be observed at a number of different length scales. One-phase flow models can capture the macroscopic flow behavior at large length scale, that is, at the length scale of mold and fiber mat. In this category, only resin flow is considered. Single-phase Darcy's law for incompressible flow is then applied to predict the resin flow behavior such as flow-front location, mold-filling time and pressure distribution inside the mold with help from continuity equations, initial conditions and boundary conditions. These models have been quite popular among researchers because they are simple enough and can fulfill the needs for engineering applications. Even without the factor of airflow that will be involved in two-phase flow models, the formulation of one-phase flow models are still very complicated and often can be expressed only in an implicit form for fairly complex mold shapes. Typical solution to this is using FEM or other numerical methods to solve the equations. The 1D flow models are proposed to further simplify the problem and provide analytical solution that can be done in a short period of time [3–6].

6.2 Key parameters and challenges

Most of the components manufactured by HLU can be manufactured by RTM process with improvements, if the proper manufacturing technology such as injection strategy can be developed. However, the major challenge for converting HLU process to RTM especially for large-scale objects is proper mold design, which involves the injection strategy, pressure and temperature distribution. Several unresolved issues in RTM encountered by composite engineers are mainly in the areas of process automation, tooling, mold flow analysis and resin chemistry. One of the major issues faced is how and from where should the resin be injected into the fiber network. The goal is to produce a void-free product with uniform distribution of the resin and to get the least possible filling time. Currently this objective is met by an iterative trial-and-error method, which results in a large development cost. All this would be a function of the fiber network, geometry and fiber volume fraction. The maximum pressure or maximum flow rate available will also determine the strategy. The resin impregnation and movement could not be seen inside a closed mold, so the only way to check the success was to wait for the part to cure. This method produced many rejects. Development of a proper manufacturing process for a complex or large-scale product requires a mold design and a successful injection strategy.

6.3 Scope and objectives

The potential of mathematical models and numerical simulations to study mold filling is immense. The transport issues to be addressed include impregnation of the resin. The fiber may be highly anisotropic and heterogeneous. The viscosity of the resin changes during curing. Hence, the flow equations may be coupled with heat transfer equations and cure kinetics to predict results. Flow modeling will allow controlling the process parameters and designing the tools accordingly. Location of the injection gates and vents and flow rates or injection pressures to be applied here can be found easily. In this context, this work has a two-fold objective. The main objective is to study and evaluate the various commercially available finite element method (FEM) software packages like ANSYS. The second is to develop an effective injection strategy, which includes location of vents and gates, max injection pressure so as to reduce the filling time and avoid formation of dry spots by performing mold filling simulations for industrial parts. The initial part of this work aims at evaluating the range of applicability of ANSYS to perform mold filling simulations. Specialized RTM packages like PAM-RTM will also be evaluated to overcome the shortcomings of the standard available packages. Mold filling simulations will be initially performed on simple objects. The latter part of the work aims at using the appropriate package to perform mold filling simulations and predict the effective injection strategy for industrially important objects that are currently being manufactured by HLU. These are shown in the figure below.

6.4 Isothermal mold filling simulations – process flow model

In the RTM process, the resin flows through a fibrous reinforcement, which can be considered as a porous medium. In this case, the flow of resin is governed by Darcy's law, which states that the flow rate of resin per unit area is proportional to the pressure gradient and inversely proportional to the viscosity of the resin. The constant of proportionality is called the permeability of the porous medium. It is independent of the fluid, but it depends on the direction of the fibers which form the reinforcement (if the porous medium is nonisotropic). The reinforcement is initially dry and the resin must fill the cavity. Capillary forces of attraction or repulsion act to the forehead of flow. These forces, which depend on the surface tension of the resin and on its ability to adhere to the surface of fibers, have the effect of either reducing or increasing the effective pressure at the resin front. However, they are considered as sufficiently small compared to the pressure field in RTM to be neglected by numerical models [7–12].

Darcy's law states that the fluid velocity is proportional to the pressure gradient:

$$\vec{V} = -\frac{K}{\mu}\vec{\nabla}P \tag{6.1}$$

where K is the permeability tensor, μ the viscosity of the resin, $\vec{V}$ the Darcy's velocity and P the pressure.

In order to preserve the balance of resin mass, the velocity field must satisfy the divergence condition,

$$\nabla.\vec{V} = 0 \tag{6.2}$$

By combining the above two equations,

$$\nabla.\left(\frac{K}{\mu}\vec{\nabla}P\right) = 0 \tag{6.3}$$

If Ω denotes the cavity and $d\Omega$ its boundary, boundary conditions are necessary to solve the problem. These conditions can be of two types:

a. Dirichlet conditions, or imposed pressure:

$$\mathrm{P} = \mathrm{f(x, y, z)} \tag{6.4}$$

This means that the pressure is specified on part of the boundary $d\Omega$. This is also the case when the injection is made under vacuum; the pressure at the inlet gate is then simply the air pressure. At the inlet gates, the pressure is equal to the value fixed by the injection pump.

b. Neumann conditions or imposed flow rate at the inlet gates:

$$\vec{V}.\vec{n} = Q \tag{6.5}$$

Resin flow simulation is performed using commercial software PAM RTM 2004. The flow is solved using a nonconforming finite element approximation. The pressure is discontinuous along the interelement boundaries except at the middle nodes for a triangular element. Contrary to conforming finite elements, the computed Darcy flow rates remain continuous across the boundary of elements. Instead of associating fill factors with the nodes of the mesh as in the conforming finite element, they are based on the elements of the mesh.

6.5 Isothermal mold filling simulations using ANSYS

The isothermal mold filling simulations through FEM package ANSYS is performed using analogous nature between flow and heat transfer process models. The correspondence between the process parameters in the heat transfer and Darcy's flow field and the use of homogenization techniques in solving Darcy flow field is well explained in the research work by Dimitrovova and Faria [13]. Heat transfer analysis was performed on a rectangular mold with injection along the smaller side. A specimen with the total length L = 0.3 m (mat permeability/resin viscosity), K/μ = 25 m^3 s/kg and mat porosity Φ = 0.5 is chosen. Injection pressure, P_0 = 5 E^{-5} Pa was used at the injection gate. The analytical filling time calculated is 18 s. On performing the analysis with a time step of 0.05 s, we obtained a numerical filling time at 17 s. The flow-front progression is shown at various time intervals in Figure 6.1.

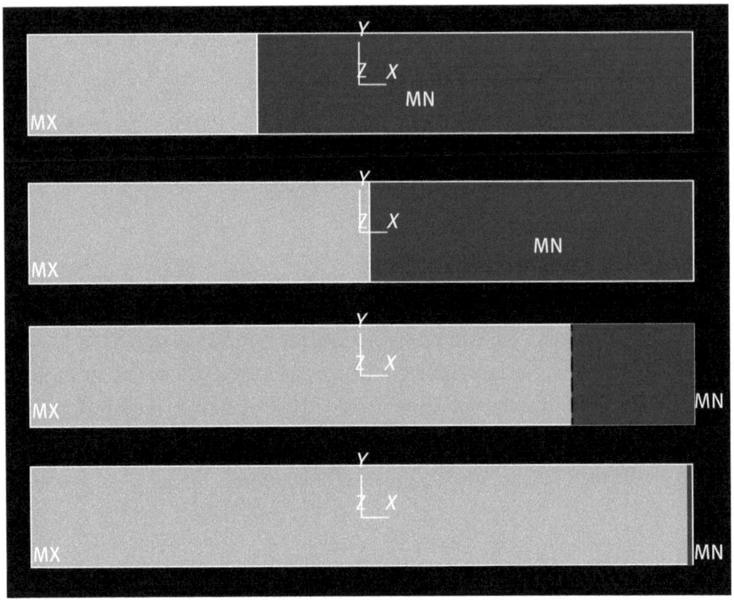

Figure 6.1: ANSYS simulated flow fronts at 1.5, 5, 10.3 and 17 s for rectangular mold.

A more realistic example was studied with a square mold of 40 cm and thickness of 4.6 mm. The resin was injected at the center through a circular inlet of diameter 6 mm. K/μ = 9 E^{-9} m^2. A constant injection flow rate of 0.4 L/min was used, which gave a filling time of 64.5 s which is very close to the experimental observed at 67 s. The run was also able to predict the maximum injection pressure of about 120 kPa. The flow-front positions are shown in Figure 6.2.

Advantages of the proposed approach with respect to the FE/CV (ANSYS) method:

- Powerful postprocessor allowing good visualization of resin velocities and front progression
- Possibility of studying a great variety of plane and shell shapes

Disadvantages of the proposed FE/CV (ANSYS) approach are:

- Slow convergence and high computer resources
- Difficult end-analysis criteria
- Complicated for nonisothermal case
- Nonplanar bodies cannot be modeled
- Difficulty in introducing direction permeability

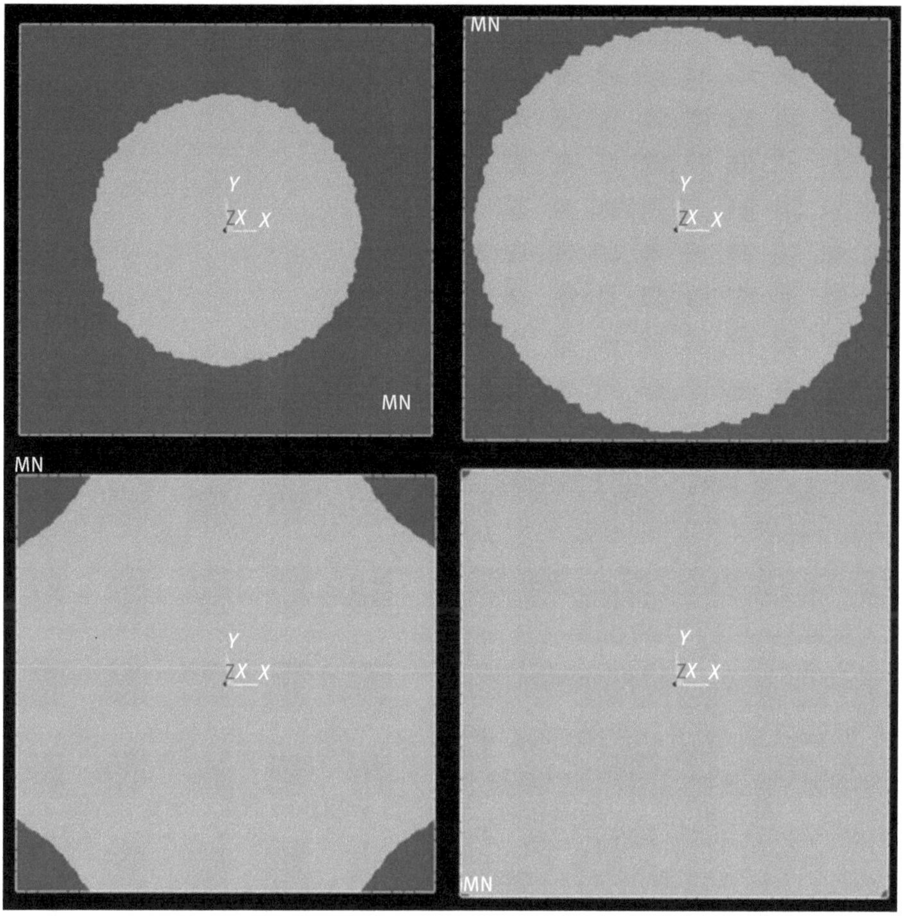

Figure 6.2: ANSYS simulated flow fronts at 22, 46, 60 and 64.5 s for a square mold.

6.6 Isothermal mold filling simulations using PAM-RTM software

The failure of the commercially available FEM packages to simulate resin flow and mold filling for complex 3D structures has given rise to the need of specialized packages dedicated solely to the RTM process. ANSYS cannot be used to simulate resin flow in a closed mold because:

- Thin objects, where the thickness is very less, are modeled using 2D planar mesh. So nonplanar objects do not produce connectivity.

- The fibers are generally arranged in layers and ANSYS cannot take care of the directional permeability for curved surfaces.
- Temperature effects cannot be implemented.

In PAM-RTM, the flow is solved using a nonconforming finite element approximation. The pressure is discontinuous along the interelement boundaries except at the middle nodes for a triangular element. Contrary to conforming finite elements, the computed Darcy flow rates remain continuous across the boundary of elements. Instead of associating fill factors with the nodes of the mesh as in the conforming finite element, they are based on the elements of the mesh.

PAM-RTM optimizes parameters that directly impact the pressure distribution during the mold filling such as injection pressure, flow rate, molding temperature, closure forces and the position of injection gates and vents. It can be effectively used to simulate flow in 3D objects. It can also predict the curing evolution inside the mold.

The test case for a square mold as explained in Section 6.5 was simulated again using PAM-RTM. The filling time was predicted to be 66.5 s, which is very close to the experimentally observed 67 s. Figure 6.3 shows the filling times and Figure 6.4 shows the pressure distribution. The maximum injection pressure calculated is 163 kPa, which is higher than that predicted by ANSYS. This may be due to the approximate nature of the analysis using ANSYS. The pressure distribution can be used to design the mold.

A nonplanar 3D part, a T-Detail, which is the substructure of a larger finbox, was used to check the suitability of PAM-RTM. The dimensions of the T-Detail are 200 mm × 200 mm × 100 mm. It has a uniform thickness of 5 mm everywhere. A fibrous mat of uniform permeability $K = 1\ E^{-9}\ m^2$ and resin of 0.1 Pa-s viscosity were implemented. Two injection strategies were implemented:

a) In the first case, a constant pressure $P = 1.5\ E^5$ Pa inlet was maintained at the top corner. The filling time was found to be 63.2 s. The flow-front progression is shown at various time intervals in Figure 6.5.
b) In the second case, a constant pressure $P = 1.5\ E^5$ Pa inlet was maintained at one bottom corner while the other three corners were maintained at atmospheric pressure. The filling time was found to be 76.4 s. The filling time is more because of gravity effects. The flow-front progression is shown at various time intervals in Figure 6.6.

6.7 Isothermal mold filling simulations – impacts

Most of the industrial composite parts that are of complex geometry are manufactured by HLU. RTM is a better substitute to it, but is not used readily due to the key challenges of mold design and unavailability of efficient injection strategies. Also

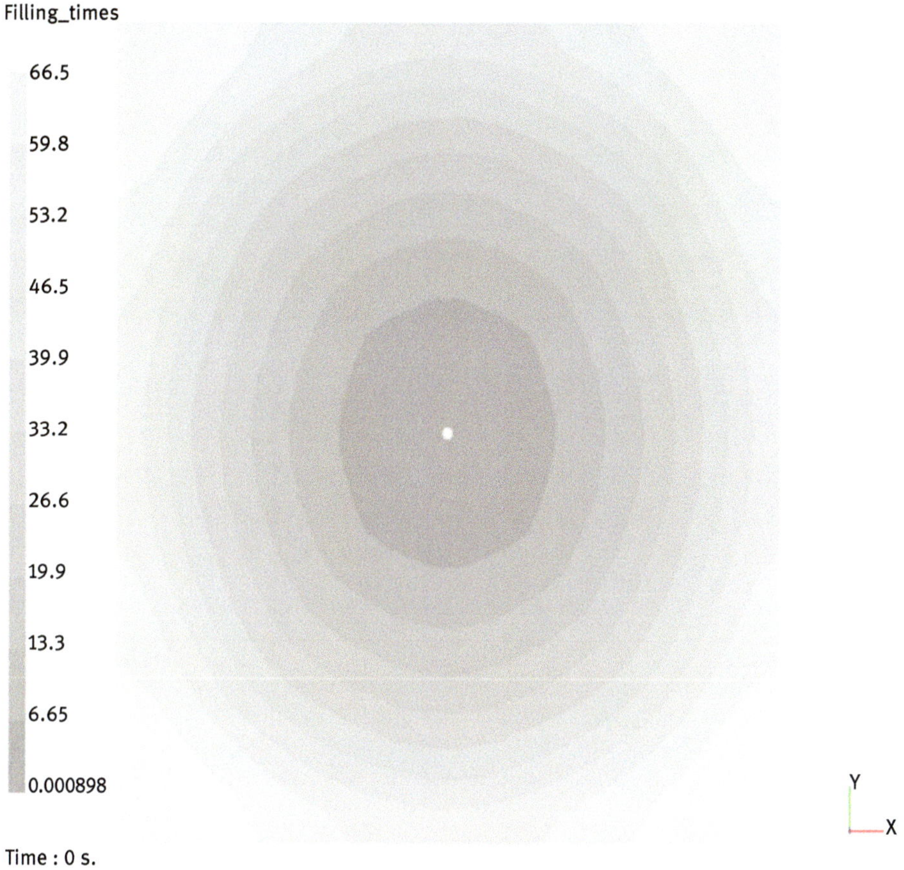

Figure 6.3: PAM-RTM simulated flow-front distribution for a square mold.

using the hit-and-trial method can prove to be very costly for large and complex objects. Mold filling simulations are performed to get an idea of the resin movement inside the mold. They can help to devise an efficient injection strategy, which will reduce filling time and injection pressure. This will in turn help to design the mold to be used. ANSYS can be used to simulate resin flow for simple objects that are not thick and can be effectively represented in 2D. Since ANSYS does not support the creation of a 2D mesh for nonplanar objects, so its use is limited. Also as the analysis uses the analogous heat transfer problem to simulate flow, the temperature effects cannot be incorporated. PAM-RTM, on the other hand, is dedicated to RTM. It can be used for a variety of shapes and it can effectively take care of direction permeability and layering. It can also be used to study curing inside the mold. The simulations performed for the test cases show that the location of inlet gates, vents and the injection pressure is very critical to the resin movement and filling time. Filling

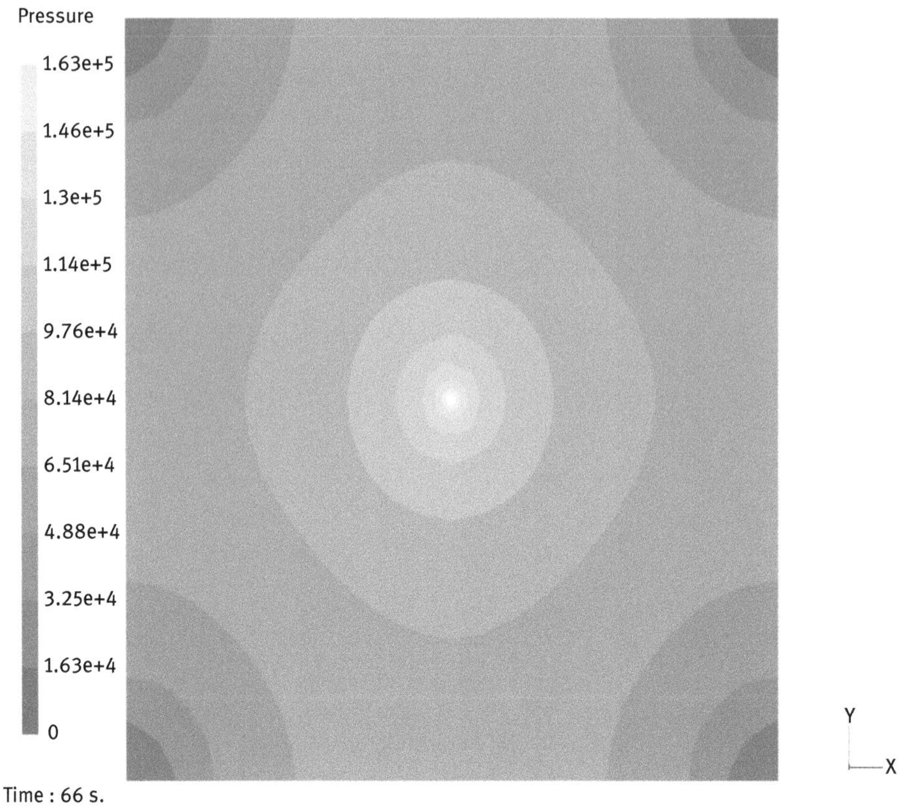

Figure 6.4: PAM-RTM simulated pressure distribution for a square mold.

time is an important factor, if it is too long the resin will cure even before filling the mold. Thus optimum filling time reduces the cycle time and also reduces dry spots.

Future work should include performing simulations on a 3D mesh for the same cab front to study the formation of possible dry spots and effects of race tracking. A 3D model can give a more realistic picture of the flow progress in the mold. After developing the most effective injection strategy, simulations can also be performed on a scaled-down model of the same object. The results will be helpful to determine the scaling down principle for a composite structure. Presently the industrial thumb rule is to use a maximum pressure of 5 atm but the modern equipment available can be used to get pressures as high as 70 atm. So an experimental verification of the mold filling with higher pressures can improve the process. Since more products that were made with HLU earlier are being manufactured by RTM, the need for mold filling simulations to predict the injection strategy becomes very important.

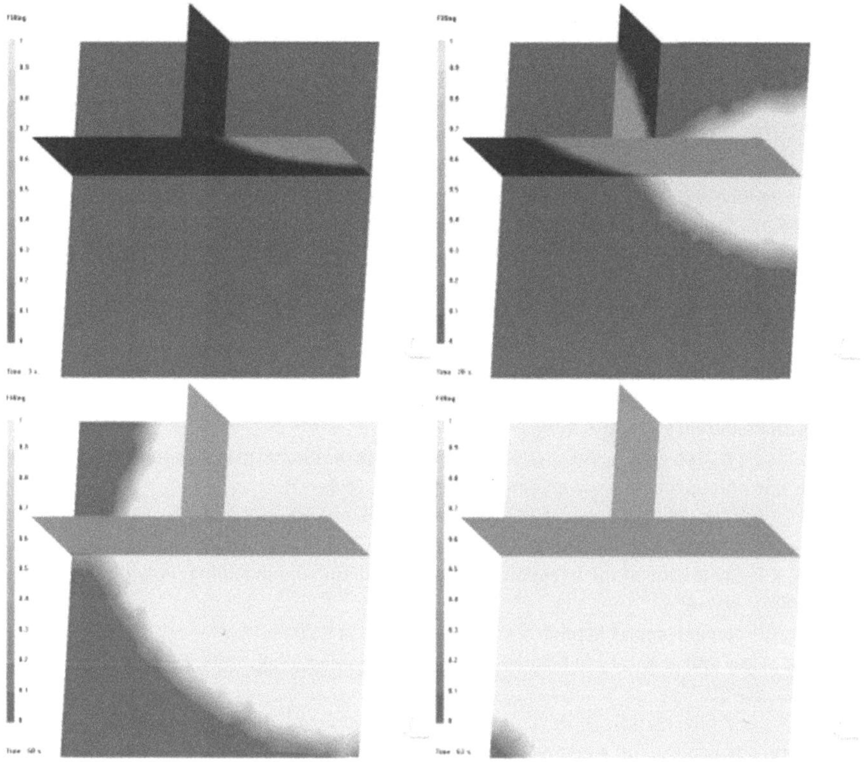

Figure 6.5: PAM-RTM simulated flow fronts at 3, 20, 50 and 63 s for T-Detail structure having an injection port at the top corner.

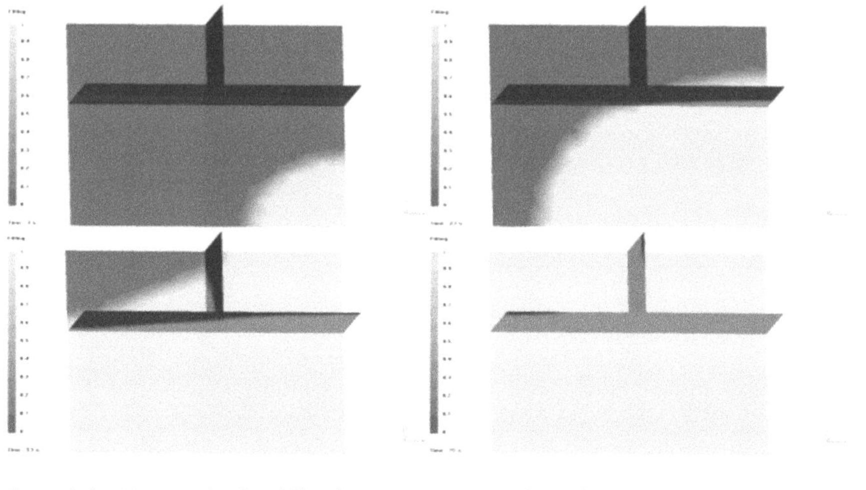

Figure 6.6: PAM-RTM simulated flow fronts at 3, 23, 53 and 75 s for T-Detail structure having an injection port at the bottom corner.

References

[1] Kuppusamy, Raghu Raja Pandiyan., Kundu, G., Neogi, S., and Patel, J. Development of manufacturing technology for cab front using resin transfer molding process. J Compos Mater 2010, 44(18), 2217–2231.

[2] Kuppusamy, Raghu Raja Pandiyan., and Neogi, S. Development of resin transfer molding process using scaling down strategy. Polym Compos 2014, 35(9), 1683–1689.

[3] Kuppusamy, Raghu Raja Pandiyan. Development of Liquid Composite Moulded Thermoset Composite Automotive Parts Using Process Simulations, Design and Optimization of Mechanical Engineering Products, Book Chapter No. 2, 2018, 24–36.

[4] Shojaeia, A., and Ghaffariana, S.R. Modeling and simulation approaches in the resin transfer moulding process: a review. Polym Compos 2003, 24(4), 525–544.

[5] Fracchia, C.A., Castro, J., and Tucker, C.L. (1989). A finite element/control volume simulation of resin transfer mould filling, Proceedings of the American Society for Composites, Fourth Technical Conference, pp. 157–166.

[6] Bruschke, M.V., and Advani, S.G. A finite element/control volume approach to mould filling in anisotropic porous media. Polym Compos 1990, 11(6), 398–405.

[7] Lim, S.T., and Lee, W.I. An analysis of the three-dimensional resin transfer mould filling process. Compos Sci Technol 2000, 60(7), 961–975.

[8] Phelan, R.F. Simulation of the injection process in resin transfer moulding. Polym Compos 1997, 18(4), 460–476.

[9] Advani, S.G. Resin Transfer Moulding Flow Phenomena in Polymeric Composites, Advani, S.G., ed., Flow and Rheology in Polymer Composite Manufacturing, 1994, Elsevier Science, Newark, 465–515.

[10] Varma, R., and Advani, S.G. Three-dimensional simulations of filling in resin transfer moulding, advances in finite element analysis in fluid dynamics (ASME). FED 1994, 200, 21–27.

[11] Lin, M., and Hahn, T. A robust and efficient approach for RTM simulation. Adv Mater 1996, 74, 75–76.

[12] Kang, M. K., Jung, J. J., and Lee, W. II. Analysis of resin transfer molding process with controlled multiple gates resin injection. Compos: Part A 2000, 31, 407–422.

[13] Dimitrovova, Z., and Faria, L. Finite element modeling of the resin transfer molding process based on homogenization techniques. Comput Struct 2000, 76, 370–397.

Wafa Ouarhim, Fatima-Zahra Semlali Aouragh Hassani, Rachid Bouhfid, Abou el kacem Qaiss

7 Numerical, experimental and simulation study of natural fiber-based composites on injection molding

Abstract: Nowadays, natural fibers are emerged in the composite materials, owing to their environment-friendly and mechanical properties. However, as reinforcement, their use in composite material should be well studied, especially during the injection molding. This chapter is dedicated to studying the effect of injection molding on the natural fiber-based composites in terms of length distribution, dispersion and orientation of fibers. Moreover, the focus is on the defects of this process on the final injected-molding part such as warpage and shrinkage. Finally, the study highlights that the final properties of the injected-molding composite are attributed to some parameters, including aspect ratio of the fibers, fiber concentration and injection parameters.

Keywords: Naturel fibers, composite properties, injection molding, orientation of fiber, warpage, shrinkage, aspect ratio, fiber concentration.

7.1 Introduction

Nowadays, an increase in the wide usage of high-tech composite materials in many sectors (transport: aeronautics, automotive, railway, leisure, sport, etc.) is closely tied to their performances and to their lower density than metal materials. Researchers' increasing willingness to control natural resources promotes the development of alternatives to materials derived from fossil resources. These involve the development of high-performance thermoplastic matrix composites and the use of natural fillers as alternatives to synthetic fibers (glass, carbon, Kevlar) as reinforcing materials [1–3]. Since injection molding provides products with high-dimensional tolerances at short cycle times and cost savings, it is an important manufacturing process for the polymer.

Wafa Ouarhim, Fatima-Zahra Semlali Aouragh Hassani, Moroccan Foundation for Advanced Science, Innovation and Research (MAScIR), Composites and Nanocomposites Center, Rabat, Morocco; Mohammed V-Rabat, University, Faculty of Science, Mechanics of Materials Laboratory, Rabat, Morocco
Rachid Bouhfid, Abou el kacem Qaiss, Moroccan Foundation for Advanced Science, Innovation and Research (MAScIR), Composites and Nanocomposites Center, Rabat, Morocco

https://doi.org/10.1515/9783110655049-007

There are other ways of making thermoplastic composites such as, for instance, compression molding. Injection remains the preferred method for making parts of complex geometries with a high rate and a low cost of production. The injection molding market is growing, along with the increasing number of application used for natural fiber composites. The potential of these combined materials with the productivity of the injection process meets the environmental challenge and industrial performance. This technique is used essentially to process unfilled and reinforced thermoplastic materials. The injection molding machine is composed of two units: an injection unit and a clamping unit. Indeed, on the one hand, the main role of the injection unit is to plasticize the material and to inject it into the mold. The screw of this unit is composed of three sections: feed, compression and metering as illustrated in Figure 7.1. Generally, the length of the feed section is the half-length of the screw, while the compression and metering sections account together the residual half-length with approximately the same length. The pitch varies between 0.8 and 1 D from the feed to the metering section across the screw flights. On the other hand, the clamping unit allows to open and close the moveable platen that holds mold, and to eject the components from the mold. As described earlier, the thermoplastic resin in molten form arrives from the injection unit to the clamping unit. Melt matrix enters into the mold from the sprue bushing, which is directly connected with the injection barrel by the nozzle (see Figure 7.2). The sprue feeds the cavity through channels termed also runners. The melt flows through the runners and arrives in the cavities by the gates. Ejection is possible by a sprue puller and ejector pins, also known as knock-out pins, located throughout the cavity and the runners. In order to keep the mold at a desired temperature to solidify the polymer, a cooling system passes water through a series of channels drilled and connected in the mold to form a continuous pathway. The manufacturers of injection molding machines play an important role in developing injection molding technology, process control, to essential mechanical engineering.

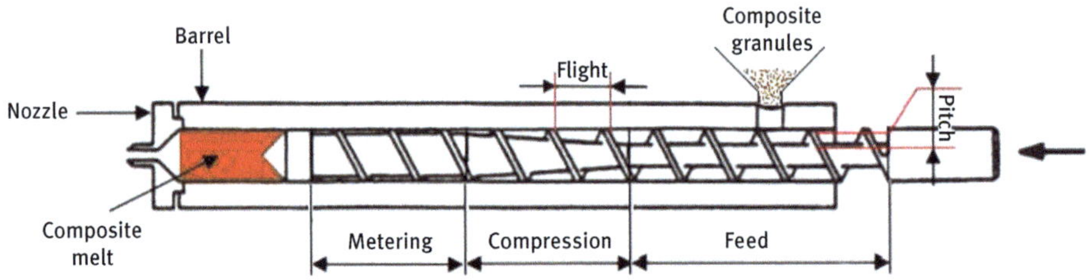

Figure 7.1: The three-section screw in the injection unit.

This chapter sheds more light on the effect of injection molding process on the composite reinforced with natural fibers in terms of fiber length and fiber dispersion and orientation. Then, the defects of the injected-molding part will be highlighted,

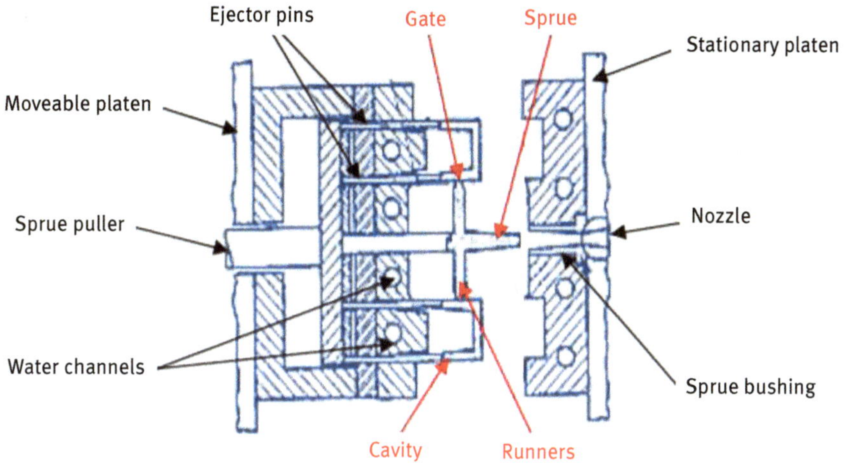

Figure 7.2: Clamping unit.

including the warpage and shrinkage. Finally, the focus is on the theoretical, experimental and simulation study of the natural fiber-based composites obtained by injection molding process.

7.2 Effect of injection molding process on the natural fiber-based composites

This section highlights the effect of injection molding process on the composites reinforced with natural fibers, including the effect of the injection process on the distribution of the natural fiber size, and on the dispersion and fiber orientation. Before attacking this, it is important to remind some basic points about the injection molding process. A typical cycle of the injection molding process includes four main phases as follows.

Plasticizing: In this step, the granules are progressively heated, melted and homogenized by the shear due to the rotation of the screw. The energy required for fusion comes from the combined action of mechanical dissipation and heat transfer from the heated sheath.

Filling: At the chosen moment, the screw is pushed axially and plays the role of a piston by injecting at a controlled speed the molten polymer into the cavity of the mold. During filling, the fibers will be oriented in privileged directions according to the applied solicitations.

Compaction: Once mold cavities have been filled, high pressure is maintained as far as possible until the material is frozen. An additional quantity of polymer is then introduced into the cavity, in order to limit the heat shrinkage and gradually standardize the pressure in the mold.

Cooling and ejecting: The room remains a few moments in the mold so that cooling can continue. Once the part is cooled, the mold opens and the part is ejected (ejectors, robot, etc.). During this time, the plasticizing phase starts again and the next cycle began.

7.2.1 Distribution analysis of the fiber length after injection

As glass fibers, natural fibers are broken down when mixed with the polymer and when injected into a mold. Therefore, studying and understanding the evolution of the fiber size during processing is extremely important. In general, the initial natural fiber size shows a decrease after processing. In the injection molding process, the plasticizing step is one of the origins for the decrease in the average length of the fibers and will thus induce a nonhomogeneous length distribution in the flow and also when flowing into the supply channels and the cavity itself [4]. Indeed, high fiber attrition can be caused by this processing technique based on several parameters, including fiber content (fiber–fiber interaction in high fiber content), viscosity of the polymer melt and melt-flow velocity [5, 6]. This attrition leads to fiber shortening during injection molding. Indeed, Barkoula et al. [7] investigate the evolution of flax fiber size during polypropylene (PP) injection molding. They found that when the fiber length is ≥0.8 mm, there is 55% decrease after injection. Some researchers compared the length of glass fibers and natural fibers by injection molding with PP [5, 8]. They found that the length of the natural fibers after injection reduces less than the glass fiber, which attributes to the fact that jute fibers contained frequent weaker regions caused by naturally developed microvoids.

Therefore, in the design of the composite, it is mandatory to take into account the fiber length because it should not neglect the main role of the fiber which fully supports the load. Thus, there is a critical length of these fibers for a good stress transfer termed critical length l_c. Indeed, below this length, the stressed fiber is removed from the matrix and the composite breaks for low loads. For longer lengths, the application of stress leads to fiber breakage (and not to loosening), resulting in higher tensile strength.

7.2.2 Natural fiber dispersion and orientation in the thermoplastic matrix

Failure of the fiber during processing greatly limits the mechanical properties of the injection molded component. Unfavorable orientation and dispersion during flow into the mold cavity may have a similar effect on the mechanical properties of the composites. The set of size, orientation and dispersion of the fibers is commonly referred to as microstructure in the field of fiber reinforced thermoplastics. In the injection molding method, the fiber dispersion and orientation are induced by the flow and were determined by image analysis using the software including Image-Pro and ImageJ.

During the injection processes, the fibers are transported by the flow, and their directions evolve according to the stress imposed by the matrix and the die walls. However, they orient themselves and show a privileged orientation [9]. During filling, the fiber movements subject, respectively, to shear and elongational flow. Indeed, in shear flow, the single fiber is animated by a periodic movement, with a Newtonian fluid, and never tends toward an equilibrium position. In elongational flow, the fiber oriented parallel or perpendicular to the direction of the flow. For example, this flow is predominant in the axis of a convergent or divergent geometry. In addition, the final orientation of the fibers into the thermoplastic matrix using injection molding influences several parameters, such as the gate position, the injection parameters (including the mold and melt temperature, the filling time and injection speed), the pseudo-plastic nature of the matrix, the aspect ratio of the reinforcement and the geometry of the mold [9]. Indeed, this latest parameter allows obtaining two cases: thin and thick parts. In thick pieces, the fiber orientation varies in the thickness and has a special structure termed core–skin structure. Another point should be noted herein is that during the injection molding process and near the walls, a solidified polymer layer is created characterized by undetermined orientation controlled by the fountain flow effect [9]. Thus, in injected molded part, there is a combination of converging and diverging flow and the fountain flow effect [10]. In literature, some works investigated the orientation of injection molded thermoplastic composites reinforced with natural fibers. Peltola et al. [11] studied the orientation of flax and hemp fibers in starch acetate through optical microscopy of polished cross section of elaborated composites; they indicate that fibers were predominantly oriented in the melt flow direction. A paper by Bledzki and Faruk [12] investigated the orientation of injection molded microcellular wood fiber in PP matrix. They emphasized a three-layer sandwich structure with different fiber orientation similar to a skin–core effect. Similarly, Vilaseca et al. [13] indicated the existence of skin–core effect with perpendicular orientation of the abaca fibers at the flow direction.

7.3 Defect of injection molding on the natural fiber-based composites

After detailed the effect of injection molding on the main parameters of the natural fiber-based composite (length, orientation, dispersion, etc.) during filling, this section focuses on the defect aspects of the natural fiber reinforced composite after injection, including warpage and shrinkage.

7.3.1 Warpage

Generally, the plastic injection molding is used to produce different shapes, in particular, the thick-and thin-walled parts. However, the production of the thin-walled parts remains very difficult because the melted plastic cannot easily fill the cavity of the mold, leading to the appearance of a problem in thin-walled part termed warpage [14]. It should be noted that the molded-injected part is thin walled when the part thickness is smaller than 1.5 mm and L/T is greater than 100 [15]. Therefore, the produced part by injection molding process does not have the desired shape and dimensions. Nevertheless, the choice of the right injection parameters and the appropriate plastic material is still the best way to prevent the warpage problem [14].

In the literature, some authors have studied how injection molding can develop this defect and how to optimize the injection parameters in order to reduce it. Indeed, Huang and Tai [16] have investigated the injected-molded thin-walled parts and have seen the influential effect over warpage. In this context, they have used C-mold Software for the analysis and Taguchi experimental method to decide the optimum values of the injection parameters. Therefore, packing pressure was found as the most important factors that affect warpage while the gate location and filling time have little effect. The same results have also been also found by Ozcelik and Sonat [14] by using ANOVA analysis in the case of a thin cell phone cover. Moreover, Barghash and Alkaabneh [17] found that warpage defect is affected by several injection molding parameters, including filling time, melt temperature, mold temperature and holding time. Likewise, injection pressure, hold pressure and cooling time have indicated significant influence on warpage [18–20]. Regarding the use of natural filler, Santos et al. [21] have analyzed experimentally and numerically the warpage of high-density polyethylene (HDPE) reinforced with short natural fibers of *Guadua angustifolia* Kunth (GAK) at four configurations: 0, 20, 30 and 40 wt%. The measures were carried out at 25 °C after 48 h of the injection molding process. The trends show the effect of reinforcements over warpage, which is also supported by Hakimian and Sulong [22]. The results show that the use of high fiber contents increases the warpage resistance of the composite.

7.3.2 Shrinkage

Like warpage, volumetric shrinkage is also a major defect of the injection molding process. It is influenced by several aspects, including processing conditions, shape, density changes, residual internal stress, flow pattern, filler content, nonuniform cooling rates and crystallinity [22–24]. The analysis of this defect is significantly important, especially for automobile applications of natural fiber-based thermoplastic composites because it has a high impact on assembly properties, product quality and production time [25].

Tan et al. [26] investigated the effects of short coir natural fiber-based PP composite over shrinkage reduction. They managed to reduce the shrinkage at 12.8% with 40 wt.% of fiber content. To do so, they were applied the following equation to calculate the shrinkage:

$$S(\%) = \frac{L_O - L}{L_O} \times 100 \tag{7.1}$$

where L_O is the side length of the mold cavity in the standard lab (mm) and L is the slide length of the specimen in the standard lab (mm). Moreover, Santos et al. [21] studied the effect of different content of short natural fiber-based composite on the volumetric shrinkage. They reached to reduce up to 58% with fiber increasing. The equation used to calculate the volume shrinkage is expressed as follows:

$$\text{Volume shrinkage} = \frac{V_c - V_s}{V_c} \times 100\% \tag{7.2}$$

where V_c is the cavity volume (mm^3) and V_s is the ejected sample volume (mm^3).

7.4 Study of natural fiber-based composites on injection molding

This section is dedicated to the investigation of composites reinforced with natural fibers on injection molding. Indeed, the theoretical investigation highlights models of molded-injected composites embedded by natural fibers during filling. Then, the focus is on the experimental study, which handles, on one hand, the influence of injection molding on the properties of the composite, and on the other hand, the influence of the final aspect ratio and concentration of fiber on the properties of the composite. Finally, the emphasis is on the simulation investigation of different tools used to predict flow progression, warpage and shortening of the produced injection molded part.

7.4.1 *Theoretical study:* modeling the flow of injection molded composites reinforced with natural fibers

Unlike thermoset matrices, thermoplastic matrices are often nonpolar resins. Moreover, since natural fibers tend to be polar and hydrophilic, they tend to absorb polar liquids

[27, 28]. Therefore, in the case of the thermoplastic composites reinforced by natural fibers, the wettability phenomenon between the fibers and the matrix is not at all verified, leading to limit the application of Darcy's law which takes this point as assumption. On the other hand, Semlali Aouragh Hassani et al. [29] developed an approach to predict the flow of molded-injected composites reinforced with natural fibers to obtain the flow front position and shape. They emphasized a three-dimensional investigation of viscous flow by solving the following Navier–Stokes equation:

$$\frac{\partial \vec{v}}{\partial x} + (\vec{v}.\vec{\nabla})\vec{v} = -\frac{1}{\rho}\vec{\nabla}\mathrm{P} + \frac{1}{\rho}\vec{f}_{\mathrm{vext}} + \upsilon\vec{\Delta}\vec{v} \tag{7.3}$$

where v denotes velocity, P is pressure, ρ and υ are the density and kinematic viscosity, respectively. To characterize the flow during filling, it is mandatory to determine the velocity expression by solving eq. (7.3). To do so, it could be necessary to put some simplified assumptions such as:

- Incompressible flow: $\mathrm{div}\,\vec{v} = \vec{0}$
- Steady flow: $\frac{\partial \vec{v}}{\partial x} = \vec{0}$
- Laminar flow $(\vec{v}.\vec{\nabla})\vec{v} = \vec{0}$
- External volume forces are neglected compared to the viscous force: $\vec{f}_{\mathrm{vext}} = \vec{0}$
- Pressure on the flow front: $P_{\mathrm{front}} = P_{\mathrm{atmosphere}}$
- Pressure at the gate: $P = P_{\mathrm{inj}}$
- Nonisothermal flow with an imposed temperature at the mold entrance.
- Symmetry with respect to the geometry median plane
- Flow along the x-axis: $\vec{v} = \vec{v}_x(y, z)$

If all the above assumptions are taken into account, eq. (7.3) of the flow can be simplified into a single Poisson equation [29]:

$$\frac{\partial P}{\partial x} = \eta\left(\frac{\partial^2 v_x}{\partial y^2} + \frac{\partial^2 v_x}{\partial z^2}\right) \tag{7.4}$$

where η is the dynamic viscosity. From eq. (7.4), the left side depends only on x and the right side depends on y and z; thus, there is only one solution to address this equation, which considers both sides constants in the whole domain whatever x, y and z may be. This allows assuming that the pressure varies linearly with the position (x) as follows:

$$\frac{\partial P}{\partial x} = -Cte \tag{7.5}$$

Thus, to determine the velocity expression, it is necessary to take into account the appropriate boundary conditions, including no slip at the wall, symmetry and the mold geometry.

7.4.2 Experimental study

7.4.2.1 Influence of the final aspect ratio and concentration of fiber on the properties of the composite

As previously described, in injection molding process, the natural fibers undergo the breakage during processing. This phenomenon is highly related to the stresses and deformations generated during mixing. Although the length of the fibers is crucial for there in forcing effect, the more important parameter is the fiber aspect ratio (length-to-diameter ratio) [30]. Unlike synthetic fibers during injection molding, the lignocellulosic fibers endured a lateral dissociation (decohesion) of the bundles or a formation of flexible fiber clusters. Consequently, since fibers divided into elementary fibers due to their bundle structure, their diameter decreases after injection molding as illustrated in Figure 7.3 [8]. It should be noted that when the reinforced fiber is already elementary before processing, diameter remains constant and their length decreases. The reduction of the fiber dimensions is due probably to friction phenomena that cause the erosion and decohesion of a bundle fiber. Kelly and Tyson [31] reported that for short-fiber composite, there is a critical fiber aspect ratio, which determines the mechanical properties of the composite. This critical aspect ratio depends on fiber intrinsic characteristics, matrix properties and quality of fiber/matrix adhesion. The good mechanical properties such as strength and toughness are obtained above this critical ratio. Moreover, the required critical aspect ratio can be lower if there is a good adhesion and an efficient interfacial stress transfer.

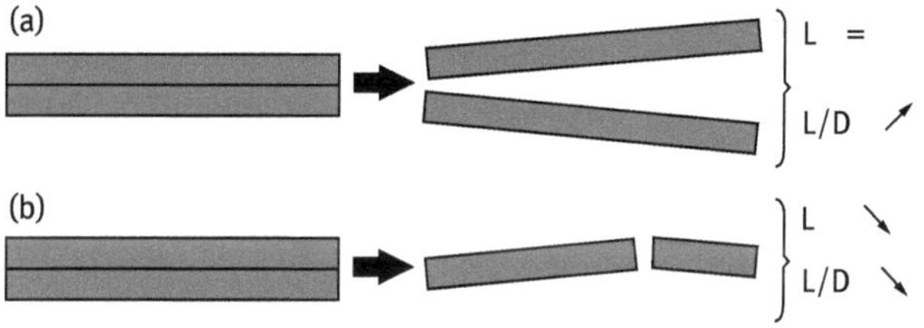

Figure 7.3: Evolution of fiber length and aspect ratio with increasing fiber concentration attributed to (a) delamination of the bundles into elementary fibers and (b) breakage of the elementary fibers into smaller fibers.

There is an approach termed masterbatch, which some authors used to produce their composite [1, 32, 33], consisting of mixing polymer with a high fiber percentage using extruder, and then the resulted composite is diluted with matrix granules directly in the injection molding step to obtain fiber percentage less than the percentage of masterbatch. It allows reducing the fiber breakage after processing because the natural fibers is undergoing less friction in the machine due to the lower time processed compared to the standard used techniques. Moreover, before the injection molding process, the composite strands coming out of the extruder pelletized into long pieces in order to reduce the probability to decrease more fiber length.

Regarding the composite properties, the high fiber concentration in the matrix leads to low mechanical properties. Indeed, Essabir et al. [34] reported a decrease of about 8–9% on tensile strength for composites reinforced with 20 wt% coir fibers, which are common for natural filler reinforced composites that act as defect point under stress (high-stress zones around fillers), leading to the formation of aggregate, especially when poor compatibility occurs. Le Duc et al. [35] highlighted that flax fibers have transverse defect termed knees or dislocations that are heterogeneously distributed over the length of the fiber. These defects depend on fiber type, growth conditions and extraction methods. As points of weakness of the natural fiber, the knees are supposed to play an important role in the localization of fracture and in the mechanical resistance. It has been verified by rheo-optics that these defects are indeed the zones of fragility where the fiber rupture is initiated. In summary, the aspect ratio of the fiber and the fiber concentration influences the properties of the molded-injected composites.

7.4.2.2 Influence of the shaping process: injection molding on the properties of the composite

Several works report on the properties of the natural fiber-reinforced thermoplastics, but only few studies correlate them with the microstructure in injection molded parts. Indeed, Aurich and Mennig [36] investigated the stiffness of 30 wt% of flax reinforced PP; the fiber aspect ratio used in this work is 13. Tensile samples were taken in parallel and perpendicular directions to the main flow directions, and elastic moduli of 3,020 and 2,200 MPa, respectively, were obtained. They explained this difference between the parallel and perpendicular elastic moduli by the different fiber alignment in the core and shell regions. Experimentally, to identify the fiber orientation in order to determine the mechanical properties of the composite, several techniques have been used for this purpose. The most widely used method for fiber orientation characterization in molded parts consists in observing a polished cross section, either by optical or scanning electronic microscopy [37]. In this context, Bourmaud et al. [38] studied the stiffness of 13.7 vol% flax-reinforced PP; they reached a difference in terms of strength and stiffness between core and shell specimens as flax fibers are more aligned in the shell than in the core. From these results, it shows that the mechanical properties of the injection molded composites vary according to fiber orientation at the matrix [39, 40]. So, in the longitudinal direction, the fibers support a high tensile strength and a low compressive strength due to fiber buckling. Moreover, the transversally aligned fibrous composites are characterized by very low tensile stress. However, in the case of randomly oriented short fiber composites, the prediction of mechanical property is very difficult, because of the fiber dispersion, orientation and complexities for the low distribution along the fiber–matrix interface [41]. Thus, by controlling parameters including aspect ratio, the dispersion and orientation of fibers, it is possible to significantly improve the properties of the composites [42].

7.4.3 Numerical simulation study

The simulation analysis is used to previously predict the injected-molded composite reinforced with natural fibers. Several commercial flow analysis software can be used for this purpose. Nowadays, software development companies providing flow analysis software have made considerable efforts to improve the technology and usability of their software. Generally, this software solves numerically the governing equation of momentum, mass and energy of injection molding shaping process [43].

In literature, the simulation software is used to compare with the various experimental parameters, including the prediction of the cavity filling phase during injection molding [29], and the simulation of the shrinkage and the warpage of injected parts [21]. Indeed, the Autodesk Moldflow Insight® software 2019 is employed by Semlali Aouragh Hassani et al. [29] to simulate the cavity filling phase of short coir fiber-based PP at different configurations (PP, PP/coir 2.5 wt%, PP/coir 5 wt%, PP/coir 10 wt% and PP/coir 15 wt%) during injection molding. The use of this software in this study was justified by the fact that it takes into account the processing conditions required during the molding injection process, including the temperature distribution, pressure distribution, melting velocity, flow front progression. They concluded that the computer simulation carried out by Autodesk Moldflow Insight® can successfully predict the flow shape for the different specimens with slight difference, which can be probably due to the surface tension, and melt shrinkage, which may be disregarded by the simulation [29]. In addition, Autodesk Moldflow® and Solid Works® are two commercial software used by Santos et al. [21] to simulate the volumetric shrinkage and warpage of molded-injected HDPE reinforced with four GAK contents (0, 20, 30 and 40 wt%). They concluded, after comparing experimental and simulation results, that both software offer similar and close results. Therefore, computer simulation remains a graceful tool to provide a bigger benefit in manufacturing time and economic profitability in the productive sector.

7.5 Conclusion

Plastic technology is one of the most dynamic manufacturing industries, characterized by new raw materials, changing requirements and continuous development of processing methods. However, the injection molding process remains the most used in this area. So, it is interesting to understand its effect on the natural fiber-based composites in order to improve their properties. Moreover, by controlling parameters, including aspect ratio, the dispersion and orientation of fibers, it is possible to significantly enhance the properties of the composites. In addition, the simulation analysis validated with experimental results is mostly used to optimize injection parameters and increase product quality. Thus, this will bring a greater benefit in manufacturing time and economic return in the productive sector.

References

[1] Ouarhim, W., Essabir, H., Bensalah, M-O., Rodrigue, D., Bouhfid, R., and Qaiss, AEK. Production and characterization of high density polyethylene reinforced by eucalyptus capsule fibers. J Bionic Eng 2018, 15, 558–566.

[2] Arrakhiz, FZ., Elachaby, M., Bouhfid, R., Vaudreuil, S., Essassi, M., and Qaiss, A. Mechanical and thermal properties of polypropylene reinforced with Alfa fiber under different chemical treatment. Mater Des 2012, 35, 318–322.

[3] Essabir, H., Raji, M., and Bouh, R. Nanoclay and natural fibers based hybrid composites: mechanical, morphological, thermal and rheological properties. Nanoclay Reinf Polym Compos Eng Mater 2016, 29–49.

[4] Tremblay, SR., Lafleur, PG., and Ait-kadi, A. Effets of injection parameters on fiber attrition and mechanical properties of polystyrene molded parts. J Inject Molding Technol 2000, 4, 1–7.

[5] Karmaker, AC., and Youngquist, JA. Injection molding of polypropylene reinforced with short jute fibers. J Appl Polym Sci 1996, 62, 1147–1151.

[6] Panthapulakkal, S., and Sain, M. Injection-molded short hemp fiber/glass fiber- reinforced polypropylene hybrid composites – mechanical, water absorption and thermal properties. J Appl Polym Sci 2007, 103, 2432–2441.

[7] Barkoula, NM., Garkhail, SK., and Peijs, T. Effect of compounding and injection molding on the mechanical properties of flax fiber polypropylene composites. J Reinf Plast Compos 2010, 29, 1366–1385.

[8] Le Moigne, N., Oever, M Van Den., and Budtova, T. A statistical analysis of fibre size and shape distribution after compounding in composites reinforced by natural fibres. Compos Part A 2011, 42, 1542–1550.

[9] Semlali Aouragh Hassani, F., Ouarhim, W., Zari, N., Bouhfid, R., and Qaiss, A el kacem. Natural fiber-based biocomposites: Effect of orientation on mechanical properties, In: Biodegradable composites: Materials, manufacturing and engineering 2019, 49–79.

[10] Semlali Aouragh Hassani, F-Z., Ouarhim, W., Bensalah, MO., Essabir, H., Rodrigue, D., Bouhfid, R. et al Mechanical properties prediction of polypropylene/short coir fibers composites using a self-consistent approach. Polym Compos 2018, 1, 1–11.

[11] Peltola, H., Madsen, B., Joffe, R., and Kalle, N. Experimental study of fiber length and orientation in injection molded natural fiber/starch acetate composites. Adv Mater Sci Eng 2011, 2011, 1–7.

[12] Bledzki, AK., and Faruk, O. Injection moulded microcellular wood fibre – Polypropylene composites. Compos Part A Appl Sci Manuf 2006, 37, 1358–1367.

[13] Vilaseca, F., Valadez-gonzalez, A., Herrera-franco, PJ., Pèlach, MÀ., Pere, J., and Mutjé, P. Bioresource technology biocomposites from abaca strands and polypropylene. Part I : Evaluation of the tensile properties. Bioresour Technol 2010, 101, 387–395.
[14] Ozcelik, B., and Sonat, I. Warpage and structural analysis of thin shell plastic in the plastic injection molding. Mater Des 2009, 30, 367–375.
[15] Chen, C., Chen, T., Chien, R., and Chen, S. Investigation on the weldline strength of thin-wall injection molded ABS parts. Int Commun Heat Mass Transf 2007, 34, 448–455.
[16] Huang, M., and Tai, C. The effective factors in the warpage problem of an injection-molded part with a thin shell feature. J Mater Process Technol 2001, 110, 1–9.
[17] Barghash, MA., and Alkaabneh, F. Shrinkage and warpage detailed analysis and optimization for the injection molding process using multistage experimental design. Qual Eng 2014, 26, 319–334.
[18] Nagahanumaiah, B. Ra. Effects of injection molding parameters on shrinkage and weight of plastic part produced by DMLS mold. Rapid Prototyp J 2009, 15, 179–186.
[19] Hajiha, H., Sain, M., and Mei, LH. Modification and characterization of hemp and sisal fibers. J Nat Fibers 2014, 11, 144–168.
[20] Chen, C., Chuang, M., Hsiao, Y., Yang, Y., and Tsai, C. Expert systems with applications simulation and experimental study in determining injection molding process parameters for thin-shell plastic parts via design of experiments analysis. Expert Syst Appl 2009, 36, 10752–10759.
[21] Santos, JD., Fajardo, JI., Cuji, AR., García, JA., Garzón, LE., and López, LM. Experimental evaluation and simulation of volumetric shrinkage and warpage on polymeric composite reinforced with short natural fibers. Front Mech Eng 2015, 10, 287–293.
[22] Hakimian, E., and Sulong, AB. Analysis of warpage and shrinkage properties of injection-molded micro gears polymer composites using numerical simulations assisted by the Taguchi method. J Mater 2012, 42, 62–71.
[23] Annicchiarico, D., and Alcock, JR. Review of factors that affect shrinkage of molded part in injection molding. Mater Manuf Process 2014, 29, 662–682.
[24] Nian, S., Wu, C., and Huang, M. Warpage control of thin-walled injection molding using local mold temperatures. Int Commun Heat Mass Transf 2015, 61, 102–110.
[25] Aizan, W., Abdul, W., Sin, LT., and Rahmat, AR. Injection moulding simulation analysis of natural fiber composite window frame. J Mater Process Technol 2008, 197, 22–30.
[26] Tan, H., Yu, Y., Xing, L., Zhao, L., and Sun, H. Polymer-plastics technology and engineering density and shrinkage of injection molded impact polypropylene copolymer/coir fiber composites density and shrinkage of injection molded impact polypropylene copolymer/coir fiber composites. Polym Plast Technol Eng 2013, 52, 257–260.
[27] Masoodi, R., and Pillai, M. K. Modeling the processing of natural fiber composites made using liquid composite molding. Handb Bioplastics Biocompos Eng Appl 2011, 43–73.
[28] Salit, MS., Jawaid, M., Bin, N., and Hoque, E. Manufacturing of Natural Fibre Reinforced Polymer Composites, Springer International Publishing, 2015.
[29] Semlali Aouragh Hassani, F., Ouarhim, W., Zari, N., Bensalah, MO., Rodrigue, D., Bouhfid, R. et al. Injection molding of short coir fiber polypropylene biocomposites : prediction of the mold filling phase. Polym Compos 2019, 1–14.
[30] Heidi, P., Bo, M., Roberts, J., and Kalle, N. The influence of biocomposite processing and composition on natural fiber length, dispersion and orientation. J Mater Sci Eng A 2011, 1, 190–198.
[31] Kelly, A., and Tyson, W. Tensile properties fibre-reinforced metals: Copper/tungsten and copper/molybdenum. J Mech Phys Solids 1965, 13, 329–350.
[32] Essabir, H., Raji, M., Essassi, EM., Rodrigue, D., and Bouhfid, R. Morphological, thermal, mechanical, electrical and magnetic properties of ABS/PA6/SBR blends with – Fe 3 O 4 nano-particles, J Mater Sci: Mater in Electr 2017, 28(22), 17120–17130.

[33] Fazal, A., and Fancey, KS. Performance enhancement of nylon kevlar fiber composites through viscoelastically generated pre-stress. Polym Polym Compos 2008, 16, 101–113.
[34] Essabir, H., Bensalah, MO., Rodrigue, D., Bouhfid, R., and Qaiss, A. Structural, mechanical and thermal properties of bio-based hybrid composites from waste coir residues: fibers and shell particles. Mech Mater 2016, 93, 134–144. doi: 10.1016/j.mechmat.2015.10.018.
[35] Duc, A Le., Vergnes, B., and Budtova, T. Polypropylene/natural fibres composites : analysis of fibre dimensions after compounding and observations of fibre rupture by rheo-optics. Compos Part A 2011, 42, 1727–1737.
[36] Aurich, T., and Mennig, G. Flow-induced fiber orientation in injection molded flax fiber reinforced polypropylene'. Polym Compos 2001, 22, 680–689.
[37] Vincent, M., Giroud, T., Clarke, A., and Eberhardt, C. Description and modeling of fiber orientation in injection molding of fiber reinforced thermoplastics. Polymer (Guildf) 2005, 46, 6719–6725.
[38] Bourmaud, A., Ausias, G., Lebrun, G., Tachon, M., and Baley, C. Observation of the structure of a composite polypropylene/flax and damage mechanisms under stress. Ind Crop Prod 2013, 43, 225–236.
[39] Godara, A., and Raabe, D. Influence of fiber orientation on global mechanical behavior and mesoscale strain localization in a short glass-fiber-reinforced epoxy polymer composite during tensile deformation investigated using digital image correlation. Compos Sci Technol 2007, 67, 2417–2427.
[40] Herrera-Franco, PJ., and Valadez-Gonzalez, A. Mechanical properties of continuous natural fibre-reinforced polymer composites. Compos Part A Appl Sci Manuf 2004, 35, 339–345.
[41] Kabir, MM., Wang, H., Aravinthan, T., Cardona, F., and Lau, K-T. Effects of natural fibre surface on composite properties : a review. Energy, Environ Sustain 2007, 94–99.
[42] Fakirov, S., and Bhattacharyya, D. Engineering Biopolymers: Homopolymers, Blends and Composites, Hanser Publications, 2007.
[43] Sahputra, IH. Comparison of Two Flow Analysis Software for Injection Moulding Tool Design, IEEE Int. Conf. Ind. Eng. Eng., 2007, p. 607–11.

Section IV: **Optimization**

Sudeepan Jayapalan

8 Optimization of injection molding process parameters for minimizing volumetric shrinkage and warpage using Moldflow simulation and Taguchi analysis technique

Abstract: With the innovations in technology, mass production of articles is facilitated by using plastic injection molding technique. However, in the manufacturing of such articles especially those with complex shapes, volumetric shrinkage and warpage play an important role for the characterization of the overall quality of plastic products in the injection molding process that may affect the accuracy of the part. The injection molding process parameters have an important influence on product quality. In this study, optimal injection molding conditions for minimizing volumetric shrinkage and warpage for rectangular-shaped tensile specimen were analyzed using Taguchi method and ANOVA technique. A model of rectangular-shaped tensile test specimen is designed and simulated using Moldflow software to imitate the real operating conditions of an injection molding process. These parameters are then examined with respect to the volumetric shrinkage and warpage occurred with the help of Taguchi method and further verified using analysis of variance technique. Confirmation tests are carried out for the optimal process parameters in order to evaluate the effectiveness of the Taguchi method. According to the results, it is found that packing pressure and melt temperature are the most significant parameters to minimize volumetric shrinkage and warpage.

Keywords: Moldflow, process parameters, injection molding, Taguchi, volumetric shrinkage

8.1 Introduction

Plastic materials are commonly used in various applications of industries such as aerospace, automotive, chemical, packaging, electronic goods and construction, because of their intrinsic material properties and the ability to change their shape as desired. The most important processing technique in the mass production of plastic products is injection molding process [1]. The process conditions, mold design and

Sudeepan Jayapalan, Department of Chemical Engineering, BIT, Mesra, India

https://doi.org/10.1515/9783110655049-008

material properties represent the high quality of injection-molded plastic products. The proper selection of process condition of injection molding is one of the most important criteria for defectless products in order to compensate volumetric shrinkage and warpage, which is common in the production of plastic products. Hence, optimization methods and design of experiments were generally applied to find out the best optimal process conditions with small number of experiments to minimize defects such as volumetric shrinkage and warpage in the products [2]. Further, simulation tools such as Moldflow and ANSYS can be used to predict the optimal process conditions to lower the manufacturing cost and time.

Many researchers have investigated the influence of injection molding process parameters on the volumetric shrinkage and warpage of different plastic products. Shen et al. [3] investigated the fiber-reinforced thermoplastic material using Moldflow analysis software by selecting process parameters such as injection pressure, melt temperature, mold temperature and injection time. Further, Li et al. [4] studied the influence of melt temperature, injection speed and injection pressure of polypropylene (PP) products using Taguchi design of experiments. Ozcelik and Sonat [5] studied the effect of process parameters on warpage and structural analysis of thin-shell plastic products. Oktem et al. [6] applied Taguchi method to minimize warpage and shrinkage by selecting process parameters, namely packing time, packing pressure, injection time and cooling time. They found that the packing pressure and packing time are the most important parameters for plastic products. Altan [7] investigated the shrinkage in injection moldings using Taguchi, ANOVA and neural network techniques. Heidari et al. [8] studied the optimization of injection molding simulation process for ultrahigh molecular weight polyethylene nanocomposite for hip liner product using response surface methodology. Further, Moayyedian et al. [9] investigated the optimization of injection molding process parameters using fuzzy logic and Taguchi experimental design. From the literature review, it is found that different researchers have used different process parameters and optimization techniques to minimize shrinkage and warpage for different plastic products.

Therefore, in this study, the influence of process parameters such as melt temperature, injection pressure, packing pressure and packing time with three different levels on the volumetric shrinkage and warpage of rectangular-shaped tensile specimen was optimized using Taguchi and ANOVA techniques. L_{27} orthogonal array (OA) design was used for conducting simulations. Simulations are carried out using Moldflow software, and Taguchi and ANOVA analyses are determined using Minitab software. Taguchi technique was used to find out the optimal conditions, and ANOVA technique is used to analyze the influence of most significant parameters on volumetric shrinkage and warpage.

8.2 Experimental details

8.2.1 Materials

In this study, commercially available homopolymer PP materials were selected. The grade of PP was injection-grade M110 manufactured from Haldia Petrochemicals, India, with a density of 0.900 g/cm^3 and melt flow index of 11 g/10 min, and its physical and mechanical properties are given in Table 8.1.

Table 8.1: Physical and mechanical properties of PP.

Property	Test method	Unit	Value
Melt flow index	ASTM D 1238	g/10 min	11
Density	ASTM D 1505	g/cm^3	0.90
Tensile strength at yield	ASTM D 638	MPa	34
Tensile elongation at yield	ASTM D 638	%	9
Flexural modulus	ASTM D 790A	MPa	1,450
Notched Izod impact strength	ASTM D 256A	J/m	30
Vicat softening point	ASTM D 1525	Degree Celsius	154
Heat deflection temperature	ASTM D 648	Degree Celsius	99

8.2.2 Model part for analysis

For the study and analysis, a rectangular-shaped tensile specimen was designed using CATIA 3D modeling software having dimensions of 190 × 19 × 3.3 mm with a gauge length of 70 mm and is shown in Figure 8.1.

8.2.3 Moldflow analysis

Moldflow MPI 6.0 software was used for the analysis to evaluate volumetric shrinkage and warpage for the model. The designed 3D CAD model was first imported into the Moldflow software by converting the 3D model into *.stl format. The feed system such as sprue, runner and gate was designed for the imported model and was meshed using fusion mesh systems. Figure 8.2 shows the meshing of the model with sprue, runner and gate system. Then the cooling channels were designed for the model and were shown in Figure 8.3. The type of analysis used for the study is Cool + Fill + Warp, and default values are selected for the injection molding machine and mold materials.

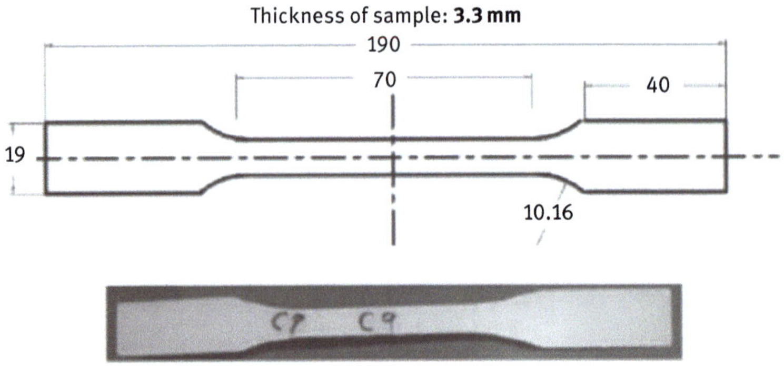

Figure 8.1: Tensile specimen model.

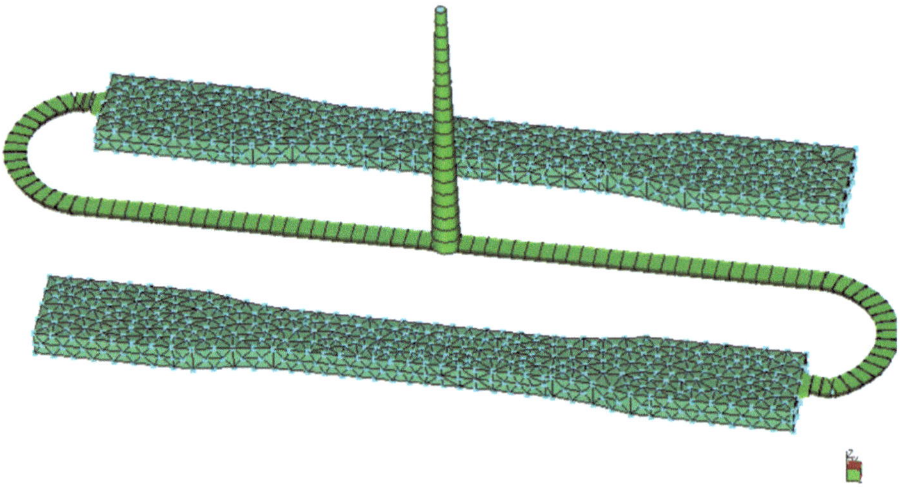

Figure 8.2: Meshing of the model.

8.2.4 Design of experiments

This study focuses on to find out the best optimum injection molding process parameter to evaluate volumetric shrinkage and warpage. The injection molding parameters such as filling time, packing pressure, melt temperature, mold temperature, packing time, gate dimensions and injection pressure play an important role in deciding the quality of molded product. The selection of most influential process parameters is one of the most important steps in the design of experiments. In this study, the process parameters selected are melt temperature, injection pressure, packing pressure and packing time with three levels of each parameter are used to determine the volumetric shrinkage and warpage of PP for the rectangular-shaped tensile specimen. The process

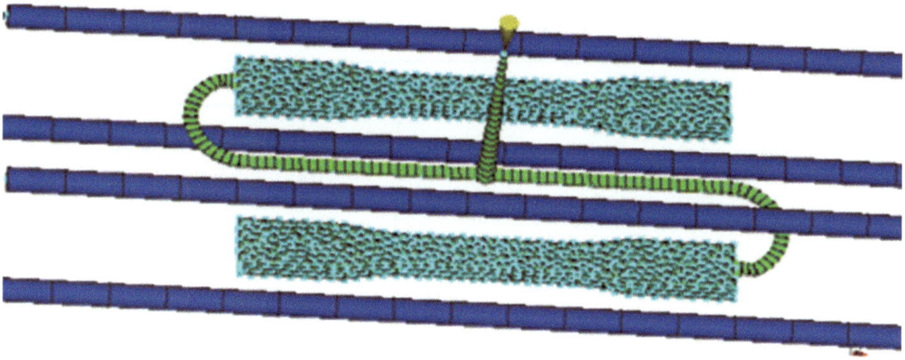

Figure 8.3: Cooling channel design.

parameters and their values that were used for the study are shown in Table 8.2. In order to determine the effect of process parameters with their levels and its interactions, predetermined OA L_{27} design of experiments are selected for this study based on total degrees of freedom (DOF) and are shown in Table 8.3. The analysis is performed as per the design of experiments given in Table 8.3 at room temperature to collect volumetric shrinkage and warpage data from Moldflow injection molding software that is commercially available.

Table 8.2: Process parameters with their levels.

Parameters	Level 1	Level 2	Level 3
Melt temperature, *A* (°C)	215	225	235
Injection pressure, *B* (MPa)	35	40	45
Packing pressure, *C* (MPa)	30	35	40
Packing time, *D* (s)	5	10	15

8.2.5 Taguchi technique

In this study, the Taguchi technique was used to study the effect of injection molding process parameters on the volumetric shrinkage and warpage. Taguchi technique is one of the most widely used statistical tools in engineering analysis to determine the high-quality performance characteristics using different design parameters with its simple and effective systematic approach [10]. Taguchi method based on OAs is used to conduct only small number of experiments to study the whole process parameters; therefore, time and cost can be reduced [11]. For engineering-related statistical analysis, use of signal-to-noise ratio (*S*/*N* ratio) has been recommended by Taguchi to determine the performance characteristics. The *S*/*N* ratio is the ratio of

Table 8.3: L_{27} OA design of experiments.

Exp. run	Melt temperature (*A*)	Injection pressure (*B*)	Packing pressure (*C*)	Packing time (*D*)
1	1	1	1	1
2	1	2	2	2
3	1	3	3	3
4	1	1	1	2
5	1	2	2	3
6	1	3	3	1
7	1	1	1	3
8	1	2	2	1
9	1	3	3	2
10	2	2	3	1
11	2	3	1	2
12	2	1	2	3
13	2	2	3	2
14	2	3	1	3
15	2	1	2	1
16	2	2	3	3
17	2	3	1	1
18	2	1	2	2
19	3	3	2	1
20	3	1	3	2
21	3	2	1	3
22	3	3	2	2
23	3	1	3	3
24	3	2	1	1
25	3	3	2	3
26	3	1	3	1
27	3	2	1	2

desired signal power to the level of background noise power [10]. The *S/N* ratio characteristics are classified into three categories: smaller-the-better, nominal-the-best and larger-the-better [12]. The aim of the study is to minimize volumetric shrinkage and warpage within the optimum level of process parameters; therefore, the smaller-the-better criterion has been selected for the study as follows:

$$S/N_S = -10\log\left(\frac{1}{n}\sum_{i=1}^{n} y_i^2\right) \tag{8.1}$$

where y represents experimental data and n denotes the number of experiments.

8.2.6 Analysis of variance

Analysis of variance (ANOVA) is a statistical technique used to determine the significant parameters on the performance characteristics. ANOVA can be carried out by separating the total variation of *S/N* ratio and comparing each parameter with the total mean of *S/N* ratio into contributions for each process parameter and the error. The percentage contribution of variance (ρ) is the ratio of sum of squared deviation (SS_d) to the total sum of squared deviation (SS_t), which is also calculated to find out the most significant parameter as follows [13]:

$$\rho = \frac{SS_d}{SS_t} \tag{8.2}$$

F-ratio is the ratio of mean sum of squared deviations also computed based on 95% confidence level to determine the significant parameter, which affects the quality characteristics. In this study, ANOVA is employed to determine the significance of each parameter in the design of experiments. Using Minitab 16 software [14], Taguchi and ANOVA analyses are performed in this study.

8.2.7 Confirmation test

Finally, a confirmation test was also carried out to validate as well as to find out accuracy of the analysis. It is calculated by comparing the initial and optimal condition results, and the estimated *S/N* ratio is computed as follows [15]:

$$\overline{y} = y_m + \sum_{i=1}^{0} (\overline{y}_i - y_m) \tag{8.3}$$

where y_m is the total mean of experimental data, $\overline{y}_i$ is the mean experimental data at the optimal testing parameter level and 0 is the number of main design process parameters that significantly affect the performance of polymer composite.

8.3 Results and discussions

8.3.1 Taguchi analysis

The simulations are carried out as per L_{27} OA (Table 8.3) based experimental design using Moldflow MPI 6.0 software, and the results of volumetric shrinkage and warpage are given in Table 8.4. As per the Taguchi technique, the simulated results are first converted to *S*/*N* ratio based on smaller-the-better criterion, which is required for the analysis as minimum volumetric shrinkage and warpage are needed for this study to determine the quality performance characteristics. In this study, Minitab software is used for converting simulated results into *S*/*N* ratio. The converted *S*/*N* ratio for volumetric shrinkage and warpage is shown in Table 8.4. Since OA is used, the effect of each parameter at different levels can be separated out to find out the mean *S*/*N* ratio for different factors. The mean *S*/*N* ratio for melt temperature, injection pressure, packing pressure and packing time for each level is computed and is shown in Tables 8.5 and 8.6 as response table for volumetric shrinkage and warpage. Based on delta value, which is the highest minus the lowest mean for each parameter, the rank of each factor is assigned and shown in the response tables. The highest delta value is ranked 1, rank 2 is assigned the second highest delta value and so on. The ranks denote which parameter is the most influential for the performance characteristics. It is found from the volumetric shrinkage response table (Table 8.5) that the parameter packing pressure which possesses rank 1 is the most significant parameter for volumetric shrinkage followed by melt temperature, packing time and injection pressure. In case of warpage (Table 8.6), it is found that the parameter melt temperature is the most effective parameter followed by packing pressure, packing time and injection pressure. From the results of *S*/*N* ratio response tables, the main effects plot for volumetric shrinkage and warpage has been drawn and shown in Figures 8.4 and 8.5 in order to find out the optimal levels of individual process parameters. As per the Taguchi analysis, the level data which is near the horizontal line has less significant effect and the one which is far away from the horizontal line will have the most significant effect on the responses. Therefore, the level with the highest *S*/*N* ratio will be the optimum process parameter for the minimum value of responses. From the main effect plot (Figure 8.4), it is seen that the optimum combination for volumetric shrinkage is found at 215 °C melt temperature, 35 MPa injection pressure, 40 MPa packing pressure and 10 s packing time (A1B1C3D2) and for warpage (Figure 8.5), it is found at 235 °C melt temperature, 40 MPa injection pressure, 40 MPa packing pressure and 10 s packing time (A3B2C3D2). It is clear from the main effect plot that the minimum volumetric shrinkage was realized at high value of packing pressure and minimum values of melt temperature and injection pressure and mid-value of packing time is required, while for minimum warpage, it is seen at high value of melt temperature and packing pressure and mid-value of injection pressure and packing time. This may be

Table 8.4: Experimental results and S/N ratio values for volumetric shrinkage and warpage.

Exp. run	Melt temperature (°C)	Injection pressure (MPa)	Packing pressure (MPa)	Packing time (s)	Volumetric shrinkage (%)	Warpage (mm)	S/N ratio (volumetric shrinkage)	S/N ratio (warpage)
1	215	35	30	5	10.85	1.16	−20.71	−1.25
2	215	40	35	10	10.38	1.14	−20.32	−1.16
3	215	45	40	15	10.10	1.14	−20.02	−1.10
4	215	35	30	10	10.79	1.15	−20.66	−1.24
5	215	40	35	15	10.38	1.15	−20.32	−1.18
6	215	45	40	5	10.09	1.13	−20.08	−1.09
7	215	35	30	15	10.79	1.15	−20.66	−1.24
8	215	40	35	5	10.46	1.14	−20.39	−1.17
9	215	45	40	10	10.07	1.13	−20.02	−1.08
10	225	40	40	5	10.40	1.10	−20.34	−0.85
11	225	45	30	10	11.09	1.12	−20.90	−0.97
12	225	35	35	15	10.67	1.11	−20.56	−0.92
13	225	40	40	10	10.29	1.10	−20.25	−0.83
14	225	45	30	15	11.09	1.12	−20.90	−0.97
15	225	35	35	5	10.78	1.11	−20.65	−0.91

(continued)

Table 8.4 (continued)

Exp. run	Melt temperature (°C)	Injection pressure (MPa)	Packing pressure (MPa)	Packing time (s)	Volumetric shrinkage (%)	Warpage (mm)	*S*/*N* ratio (volumetric shrinkage)	*S*/*N* ratio (warpage)
16	225	40	40	15	10.30	1.10	−20.26	−0.83
17	225	45	30	5	11.20	1.12	−20.98	−0.98
18	225	35	35	10	10.67	1.11	−20.56	−0.90
19	235	45	35	5	11.11	1.10	−20.91	−0.68
20	235	35	40	10	10.54	1.07	−20.46	−0.58
21	235	40	30	15	11.40	1.19	−21.14	−0.72
22	235	45	35	10	10.94	1.08	−20.78	−0.65
23	235	35	40	15	10.54	1.07	−20.46	−0.58
24	235	40	30	5	11.58	1.14	−21.27	−0.72
25	235	45	35	15	10.94	1.04	−20.78	−0.65
26	235	35	40	5	10.73	1.07	−20.61	−0.60
27	235	40	30	10	11.40	1.09	−21.14	−0.72

Table 8.5: Response table for volumetric shrinkage.

Level	Melt temperature (°C)	Injection pressure (MPa)	Packing pressure (MPa)	Packing time (s)
1	−20.35	−20.59	−20.93	−20.66
2	−20.6	−20.6	−20.59	−20.57
3	−20.84	−20.6	−20.28	−20.57
Delta	0.49	0.01	0.65	0.1
Rank	2	4	1	3

Table 8.6: Response table for warpage.

Level	Melt temperature (°C)	Injection pressure (MPa)	Packing pressure (MPa)	Packing time (s)
1	−1.1675	−0.9127	−0.9774	−0.9169
2	−0.9063	−0.9078	−0.9135	−0.9029
3	−0.6549	−0.908	−0.8377	−0.9089
Delta	0.5126	0.0049	0.1398	0.014
Rank	1	4	2	3

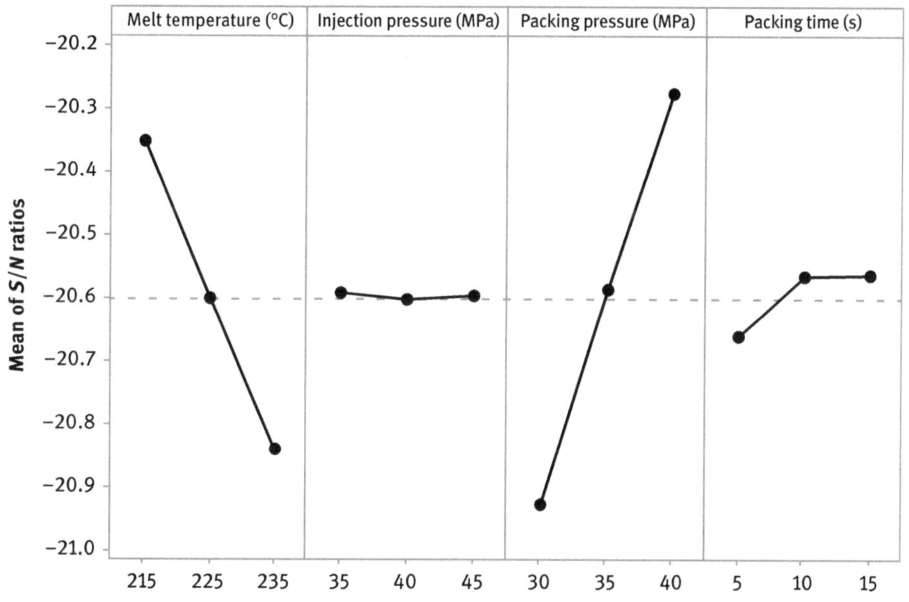

Figure 8.4: Main effect plot for volumetric shrinkage.

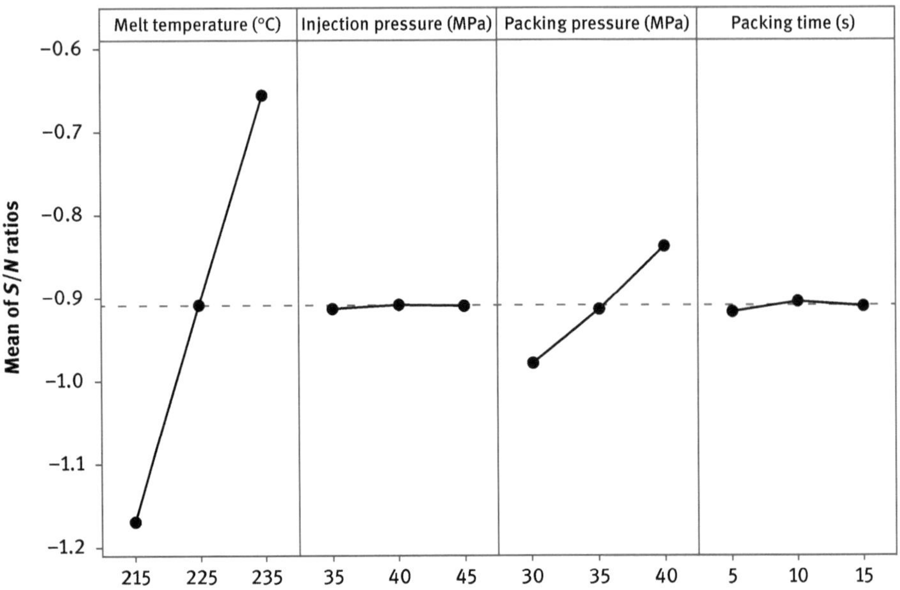

Figure 8.5: Main effect plot for warpage.

due to the cause that the high packing pressure will compensate the volumetric shrinkage during cooling of the product; hence, higher value of packing pressure is required for minimum volumetric shrinkage [7]. Further, high melt temperatures reduce internal stresses in the part; thus, high value of melt temperature is required to achieve minimum warpage [5].

8.3.2 Analysis of variance

ANOVA was carried out for main factors based on 95% confidence level to determine the significance of each process parameter on the quality characteristics of volumetric shrinkage and warpage. The percentage contribution of each process parameter was also calculated using ANOVA results. The ANOVA results for volumetric shrinkage and warpage are given in Tables 8.7 and 8.8, respectively. From the volumetric shrinkage and warpage ANOVA table, it is seen that *F*-ratio for factors such as melt temperature, packing pressure and packing time are greater than $F_{0.05,2,18} = 3.55$ has significant effect on the volumetric shrinkage and warpage. It was found from Tables 8.7 and 8.8 that the most important parameter for minimum volumetric shrinkage in case of percentage contribution (%) was packing pressure (62.96%) followed by melt temperature (34.79%), packing time (1.84%) and injection pressure (0.06%), respectively. In case of warpage, the most significant parameter is melt temperature (93.01%) followed by packing pressure (6.94%), packing

Table 8.7: ANOVA table for volumetric shrinkage.

Source	DF	Adj. SS	Adj. MS	*F*-ratio	% contribution
Melt temperature (°C)	2	1.62	0.81	890.23	34.79
Injection pressure (MPa)	2	0.01	0.02	1.64	0.06
Packing pressure (MPa)	2	2.93	1.47	1,611.27	62.96
Packing time (s)	2	0.09	0.04	47.04	1.84
Error	18	0.02	0.01		0.35
Total	26	4.66			100.00
R^2 = 99.65%; R^2 (adj.) = 99.49%; R^2 (pred.) = 99.21%					

Table 8.8: ANOVA table for volumetric warpage.

Source	DF	Adj. SS	Adj. MS	*F*	% contribution
Melt temperature (°C)	2	0.0193	0.0097	10,447.00	93.01
Injection pressure (MPa)	2	0.0004	0.0003	2.56	0.02
Packing pressure (MPa)	2	0.0014	0.0007	779.32	6.93
Packing time (s)	2	0.0005	0.0004	7.72	0.06
Error	18	0.0004	0.0003		0.08
Total	26	0.0208			100.00
R^2 = 99.92%; R^2 (adj.) = 99.88%; R^2 (pred.) = 99.82%					

time (0.07%) and injection pressure (0.02%), respectively. Also, the percentage errors for volumetric shrinkage and warpage are 0.35% and 0.08%, respectively, and *R*-square values of 99.65% and 99.92% indicate the predicted outcome of this study, which may be acceptable. Mamat et al. [16], Pontes and Pouzada [17] and Altan [7] investigated the effect of process parameters on shrinkage in injection moldings and found that the packing pressure was the most significant factor for achieving minimum shrinkage. Further, Erzurumlu and Ozcelik [11] andOzcelik and Sonat [5] investigated the warpage in injection-molded thermoplastic parts and found that the melt temperature and packing pressure are the most influential parameter to reduce warpage. In this study, similar to the previous literatures, packing pressure is the most significant parameter on the volumetric shrinkage apparently due to very close packing and shrinkage compensation on the injection moldings. In case of warpage, melt temperature is found to be the most significant parameter probably due to better pressure transmission on the moldings.

8.3.3 Confirmation test

Confirmation test was carried out to validate the minimum responses by comparing the initial and final optimal parameters as per eq. (8.3). Table 8.9 shows the values of confirmation test using initial parameter with optimal parameter. The

Table 8.9: Confirmation result table for volumetric warpage and warpage.

	Initial parameter	Predicted	Experimental
Level	A1B1C1D1	A1B1C3D2	A1B1C3D2
Volumetric shrinkage	10.85	9.98	10.02
Level	A1B1C1D1	A3B2C3D2	A3B2C3D2
Warpage	1.155	1.067	1.0860

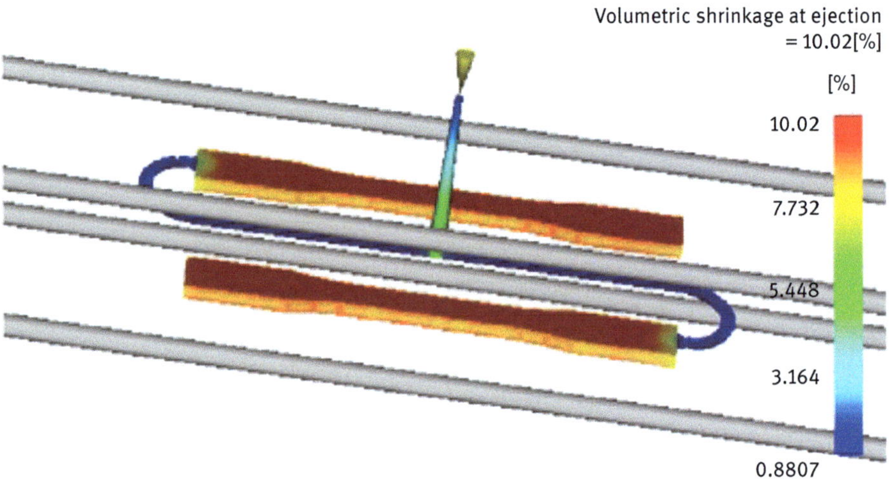

Figure 8.6: Confirmation test result for volumetric shrinkage.

improvement observed from initial to optimal was found to be 7.65% and 5.51% for volumetric shrinkage and warpage analysis, respectively. The simulated results of volumetric shrinkage and warpage for the optimal conditions are given in Figures 8.6 and 8.7.

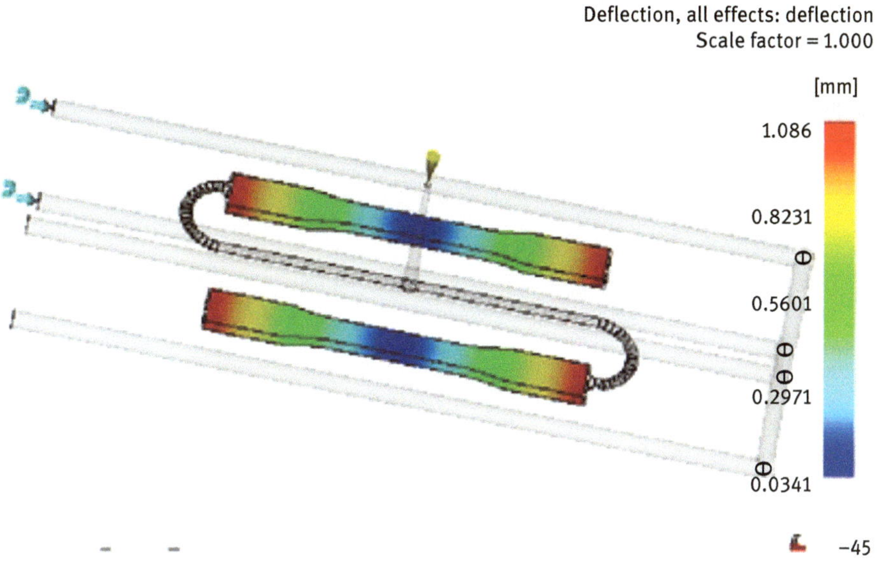

Figure 8.7: Confirmation test result for warpage.

8.4 Conclusions

In this study, Taguchi and ANOVA techniques were used to investigate the influence of melt temperature, injection pressure, packing pressure and packing time on the volumetric shrinkage and warpage of the rectangular-shaped tensile specimen. Taguchi method which converts the simulated results to *S*/*N* ratio was used to find out the optimal process parameters to determine minimum volumetric shrinkage and warpage. The response table results showed that the packing pressure is the most significant parameter for volumetric shrinkage and melt temperature is the most significant for warpage. From the main effect *S*/*N* ratio plot, it is seen that 215 °C melt temperature, 35 MPa injection pressure, 40 MPa packing pressure and 10 s packing time will be the optimum combination for volumetric shrinkage. In case of warpage, it is 235 °C melt temperature, 40 MPa injection pressure, 40 MPa packing pressure and 10 s packing time. Further, ANOVA technique was used to find out the significant parameter for volumetric shrinkage and warpage. According to ANOVA results, it is observed that the packing pressure of percentage contribution of 62.96% is the most significant factor to minimize volumetric shrinkage followed by melt temperature, packing time and injection pressure. For warpage, melt temperature having percentage contribution of 93.01% was found to be the most influential parameter followed by packing pressure, packing time and injection pressure, respectively. Finally, a confirmation test was carried out to evaluate the effectiveness of Taguchi optimization technique. It is observed that the improvement from initial to optimal parameters was 7.65% and 5.51% for volumetric shrinkage and warpage analysis, respectively. It is concluded from this study that the parameter packing pressure and melt temperature are the most effective parameters for reducing volumetric shrinkage and warpage.

References

[1] Liao, S. J., Chang, D. Y., Chen, H. J., Tsou, L. S., Ho, J. R., Yau, H. T., and Su, Y. C. Optimal process conditions of shrinkage and warpage of thin-wall parts. Polym Eng Sci 2004, 44(5), 917–928.

[2] Ross, P. J., and Ross, P. J. Taguchi techniques for quality engineering: loss function, orthogonal experiments, parameter and tolerance design (No. TS156 R12), 1988, McGraw-Hill, New York.

[3] Shen, Y. K., Yeh, P. H., and Wu, J. S. Numerical simulation for thin wall injection molding of fiber-reinforced thermoplastics. Int Commun Heat Mass Transfer 2001, 28(8), 1035–1042.

[4] Li, H., Guo, Z., and Li, D. Reducing the effects of weldlines on appearance of plastic products by Taguchi experimental method. Int J Adv Manuf Technol 2007, 32(9–10), 927–931.

[5] Ozcelik, B., and Sonat, I. Warpage and structural analysis of thin shell plastic in the plastic injection molding. Mater Des 2009, 30(2), 367–375.

[6] Oktem, H., Erzurumlu, T., and Uzman, I. Application of Taguchi optimization technique in determining plastic injection molding process parameters for a thin-shell part. Mater Des 2007, 28(4), 1271–1278.

[7] Altan, M. Reducing shrinkage in injection moldings via the Taguchi, ANOVA and neural network methods. Mater Des 2010, 31(1), 599–604.

[8] Heidari, B. S., Davachi, S. M., Moghaddam, A. H., Seyfi, J., Hejazi, I., Sahraeian, R., and Rashedi, H. Optimization simulated injection molding process for ultrahigh molecular weight polyethylene nanocomposite hip liner using response surface methodology and simulation of mechanical behavior. J Mech Behav Biomed Mater 2018, 81, 95–105.

[9] Moayyedian, M., Abhary, K., and Marian, R. Optimization of injection molding process based on fuzzy quality evaluation and Taguchi experimental design. CIRP J Manuf Sci Technol 2018, 21, 150–160.

[10] Taguchi, G. Robust technology development. Mech Eng-CIME 1993, 115(3), 60–63.

[11] Erzurumlu, T., and Ozcelik, B. Minimization of warpage and sink index in injection-molded thermoplastic parts using Taguchi optimization method. Mater Des 2006, 27(10), 853–861.

[12] Phadke, M. S. Quality Engineering Using Robust Design, 1989, PTR Prentice-Hall. Inc, Englewood Cliffs, NJ.

[13] Lindman, H. R. Analysis of variance in experimental design, 2012, Springer Science & Business Media, New York.

[14] Minitab, M. U. Making data analysis easier, 2001, MINITAB Inc, USA.

[15] Roy Ranjit, K. A primer on Taguchi method, 1990, Van Nostrad Reinhold, New York.

[16] Mamat, A., Trochu, F., and Sanschagrin, B.. (1994). Shrinkage analysis of injection molded polypropylene parts. In ANTEC, Conference Proceedings (Vol. 11, pp. 513–517).

[17] Pontes, A. J., and Pouzada, A. S. Predicting shrinkage in semi-crystalline injection mouldings – the influence of pressure, Materials science forum, 2006, Vol. 514, Trans Tech Publications, 1501–1505.

Bappa Acherjee

9 Process overview, experimental study and Taguchi quality loss function analysis of laser transmission welding of thermoplastics

Abstract: Due to extensive use of plastic products, joining methods play a significant role in their processing. The overall benefits of welding methods over other joining methods are fast and simple processing, and higher joint strength. Laser transmission welding offers an attractive alternative for achieving the limits of conventional joining methods for plastics. This chapter provides a brief overview of the laser transmission welding process. The emphasis is given on the parameters governing the welding process and the main phenomena affecting the joint performance. Experimental investigation is carried out to study the effects of parameters on the quality features of laser transmission welding. Finally, the Taguchi quality loss function is implemented to acquire the desired quality features, by realizing the optimum settings of welding parameters.

Keywords: Joining of polymers, Thermoplastics, laser transmission welding, Parametric Analyses, Taguchi Quality Loss Function Analysis

9.1 Introduction

Nowadays, plastics are found nearly everywhere, from household products to the canopies of aircraft. They are used in multiple sectors, ranging from household goods, to packaging, biomedical, buildings, electrical equipment, electronics, aerospace and automobiles [1, 2]. Because of their excellent strength-to-weight ratio, easy manufacturing of complicated shapes, low price and easy recycling, the use of plastics has grown significantly. Since it is not always practical and cost-effective to make the complicated plastic parts in one-piece, different joining methods have been used. Welding has considerable process benefits over other methods of joining, including plastic adhesive joining and mechanical fastening. The benefits of welding are quick and simple processing, joint tightness and high strength. Hot tool welding, ultrasonic and vibration welding are the most commonly used plastic

Bappa Acherjee, Department of Production Engineering, Birla Institute of Technology, Ranchi, India

https://doi.org/10.1515/9783110655049-009

welding processes. Laser transmission welding provides an appealing solution to reaching the boundaries of standard plastic joining techniques [3].

There are two broad categories of plastics: thermosets and thermoplastics. During processing (manufacturing), thermosets undergo chemical reaction that parts once formed cannot be reshaped except by machining. Unlike thermosets, without changing the chemical structure, thermoplastics can be melted and molded. Thus, thermoplastics can be welded, but thermosets cannot be welded [4]. Most plastics in their natural state are either transparent or translucent. Plastics are colored for the majority of consumer applications. The ease with which a molded part incorporates color is a benefit of plastics over metals and ceramics, which rely on color coatings. For laser transmission welding, plastic part is made to absorb laser by adding pigments of color such as carbon black.

9.2 Process overview

In laser transmission welding, a laser beam is used to weld at an interface of two overlapping plastic parts by transmitting through the top transparent part and absorbing at the bottom laser absorbing plastic part. The laser energy is absorbed over a depth in the bottom part based on the absorption coefficient and thickness of that part. The laser energy absorbed in the opaque thermoplastic in the near-infrared spectrum creates vibration of electron bonds, which causes the material to heat locally [5]. The heat generated in the bottom part is conducted within that parts and is also transferred by conduction to the top part. To keep both the parts in a place to guarantee appropriate contact with the melt for the conductive heating of the transparent material and to restrict the thermal expansion, clamping pressure must be applied on the joint. The thermoplastic enters a liquid/semiliquid state when heated to temperatures above the melting point or melting range. The polymer chains may combine by mixing or diffusing. A strong link is created at the weld seam after cooling and re-solidification. A firm connection at the weld seam is formed after cooling and re-solidification. Figure 9.1 represents the working principle of laser transmission welding of thermoplastics.

Laser transmission welding provides explicit process benefits over other welding methods, such as short cycle times while offering optically and qualitatively superior weld seams, and imposes minimal thermal stress [6]. In addition, laser transmission welding is versatile, gentle and ecofriendly. Fiber, Nd:YAG and diode lasers work in near-infrared spectrum, where plastics have negligible inherent absorption, allowing effective laser transmission through plastic components with millimeter thickness to be welded. A lap joint is the most fundamental joint setup for laser transmission welding of plastics. Whatever sort of joint is selected, there are two criteria that must be fulfilled. The first criterion is that laser beam must reach the joint interface and the second criterion is that pressure must be applied to the joint [4].

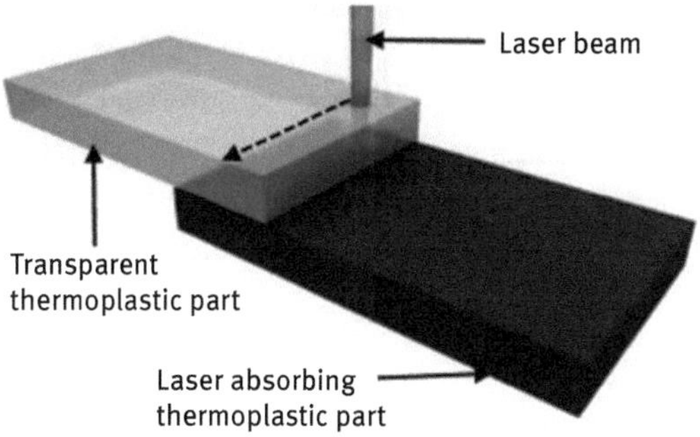

Figure 9.1: The working principle of laser transmission welding of thermoplastics.

It is possible to identify four variants of laser transmission welding strategies, namely contour, simultaneous, quasi-simultaneous and mask welding. In contour welding, a laser beam moves through the targeted region to produce a continuous welding seam. Laser beam scans a predefined weld line. In simultaneous laser welding, laser beam of the proper spot size or multiple modules of diode lasers are used to irradiate the whole weld area simultaneously, without having to move relatively between the laser beam and the workpiece. In quasi-simultaneous laser welding, with a galvomirror system, a laser beam scans the workpiece several times targeting the weld path, very quickly. The welding track heats up gradually and approximately evenly due to the low thermal conductivity of polymers, so that the entire welding track is melted quasi-simultaneously. Mask welding is a method of laser scanning a region with a mask, ensuring that only the chosen areas can be heated when the laser moves. Masks consist of steel or other materials that efficiently block laser radiation [7].

The polymer composition, additives and colorants influence the reflection, transmission and absorption of laser energy and finally the quality of the weld [8]. Plastic part thickness also has an impact on optical properties, particularly for crystalline and semicrystalline materials. Due to enhanced light scattering, the laser transmission reduces with growing fiberglass content in polymer [9]. Thus, increasing fiberglass content improves the strength of the base material but decreases at weld seam [10]. Increasing the content of fiberglass in polymer raises the minimum energy requirement during welding [11]. Natural plastics absorb a very tiny part of laser energy at near-infrared spectral wavelength [12]. Plastics pigmented with carbon black do not transmit laser energy at the wavelength of Nd:YAG, diode or fiber laser [13]. Requirement of laser energy for welding is strongly influenced by the carbon black content within the absorbing plastic part [14, 15]. The white sample (pigmented with titanium oxide) does not absorb the radiation at near infrared wavelength but scatters it to such a degree that the welding time is always drastically impacted [11]. Increasing the size of the laser beam results in wider welding width and increased welding strength [2]. With the suitable combination of laser energy and welding velocity, optimum welding strength can be accomplished [16]. Increasing the air gap between the parts decreases the tensile strength, but with roughly the same line energy as in trials without an air gap, the highest weld strength is obtained [2]. The main laser transmission welding parameters are laser power, welding velocity, beam area and clamp pressure, which regulate the temperature field within the weld and therefore the weld quality.

9.3 Experimental study

Natural transparent polycarbonate and opaque polycarbonate (added 0.1% carbon black pigments by weight proportion) are work materials used for performing experimental work. Although polycarbonate sheets are valued for their strength, flexibility, high impact and high heat resistance, their transparency enables them to take on a range of applications. For this experimental study, the contour welding configuration of laser transmission welding is adopted. To carry out the experimental investigation, the following laser transmission welding parameters are identified: laser power (A), welding velocity (B), stand-off distance (C) and clamp pressure (D). A 30-W diode laser system integrated with CNC worktable is used to perform the welding operations. A hydraulic clamping device is used to apply clamping pressure during welding. Trial runs are carried out and the operating range is determined by checking the weld seam appearance with no noticeable flaws. Table 9.1 shows the chosen welding parameters and their limits, units and symbols. Taguchi's methodology for four-level four factors is used to plan the orthogonal array experiment scheme. Experiments are conducted according to the experimental plan. The welded samples are tested using an Instron® universal testing machine for testing their strength. For measuring weld seam widths, an Olympus STM 6 measuring microscope is used. The experimental design and experimental outcomes for the welding parameters, using the L16 orthogonal array, are shown in Table 9.2. Figure 9.2 shows a photograph of a welded specimen.

Table 9.1: Welding parameters and their levels for experiments.

Symbol	Parameter	Level			
		1	2	3	4
A	Power (W)	12	14	16	18
B	Welding speed (mm/s)	7	9	11	13
C	Stand-off distance (mm)	28	30	32	34
D	Clamp pressure (MPa)	2	3	4	5

Table 9.2: Experimental layout and results of laser welding of polycarbonates.

Experiment	Welding parameters				Weld strength	Weld width
no.	*A* (W)	*B* (mm/s)	*C* (mm)	*D* (MPa)	(N/mm)	(mm)
1	12	7	28	2	38.27	2.73
2	12	9	30	3	42.66	2.67
3	12	11	32	4	41.39	2.65
4	12	13	34	5	34.47	2.67
5	14	7	30	4	50.58	3.05
6	14	9	28	5	40.61	2.62
7	14	11	34	2	48.01	3.11
8	14	13	32	3	41.69	2.69
9	16	7	32	5	60.63	3.45
10	16	9	34	4	63.18	3.37
11	16	11	28	3	43.21	2.49
12	16	13	30	2	42.19	2.51
13	18	7	34	3	74.96	3.66
14	18	9	32	2	62.69	3.06
15	18	11	30	5	52.67	2.73
16	18	13	28	4	43.28	2.31

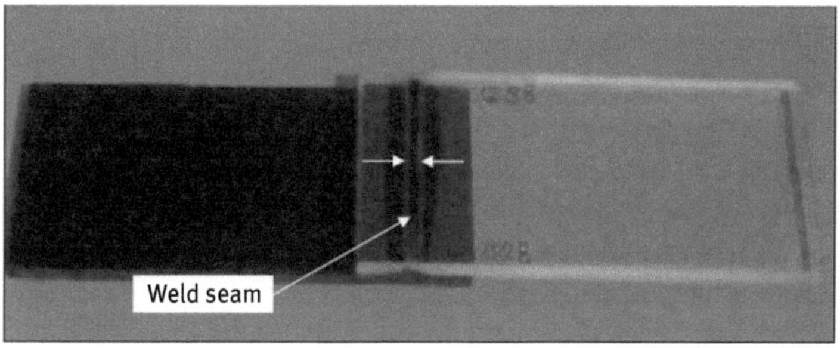

Figure 9.2: Photographic view of the weld seam.

The aim of analysis of variance (ANOVA) is to explore which welding parameters influence the quality features considerably. The results of ANOVA provided in Tables 9.3 and 9.4 are obtained using MINITAB® 17 statistical software. In ANOVA, Fisher's ratio (*F*-value) is used to determine whether the parameter has a significant effect on the selected weld quality. The associated *p*-value of less than 0.05 indicates that the parameters are statistically significant at a confidence level of 95% [17]. It can be seen from the ANOVA tables (Tables 9.3 and 9.4) that all process parameters, except clamp pressure, are statistically significant. According to ANOVA results given in Table 9.3, laser power shows the leading effect on total variation in weld strength and is followed by welding speed, stand-off distance and clamp pressure. It is observed

Table 9.3: Analysis of variance for weld strength.

Sources	Sum of squares	Degrees of freedom	Mean squares	*F*-value	*p*-Value
A	837.73	3	279.243	97.42	0.002
B	568.66	3	189.552	66.13	0.003
C	427.96	3	142.653	49.77	0.005
D	31.71	3	10.569	3.69	0.156
Error	8.60	3	2.866		
Total	1,874.65	15			

Table 9.4: Analysis of variance for weld width.

Sources	Sum of squares	Degrees of freedom	Mean squares	*F*-value	*p*-Value
A	0.1915	3	0.0638	16.92	0.022
B	0.9950	3	0.3317	87.91	0.002
C	0.9849	3	0.3283	87.01	0.002
D	0.0026	3	0.0009	0.23	0.873
Error	0.0113	3	0.0038		
Total	2.1853	15			

from ANOVA results furnished in Table 9.4 that welding speed and stand-off distance have almost the same impact, leading the source variation for weld width, which is followed by laser power and clamp pressure.

Figures 9.3 and 9.4, respectively, present the main effects plot for weld strength and weld width, and effects of laser welding parameters on those quality features. Both weld strength and weld width are observed to improve with laser power. Welding speed shows to have an adverse impact on both weld strength and weld width. Stand-off distance, which regulates the energy density over irradiation area by controlling the area of the laser beam spot on the work surface, also had a very significant effect on the strength and the width of the weld. Although the clamp pressure is statistically insignificant, clamp pressure is required to be applied to maintain close contact with the mating surfaces and to promote the melt flow within the weld pool. However, very high clamp pressure causes the ejection of molten material from weld pool through the end of the weld and may cause melt down.

9.4 Taguchi quality loss function analysis

The loss of quality is the loss owing to the deviation in the functional output of the product from its product-related objective. The experimental values in all experimental runs are used to calculate the quality loss values for each quality characteristics. The

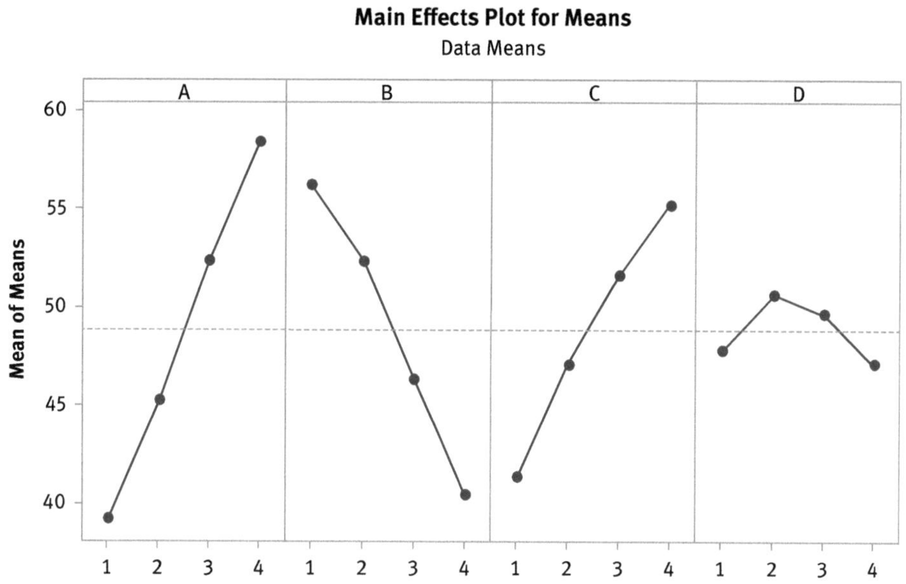

Figure 9.3: Main effects plot of welding parameters on weld strength.

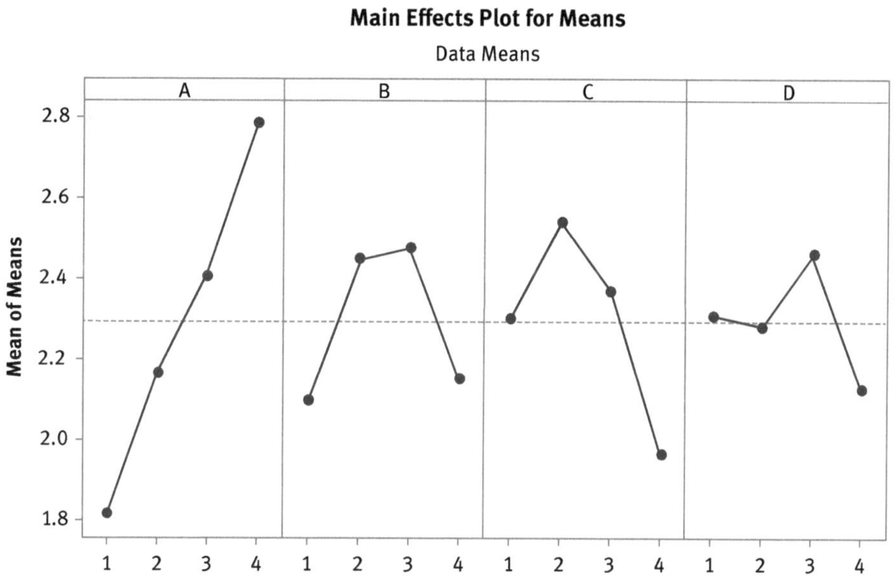

Figure 9.4: Main effects plot of welding parameters on weld width.

quality loss function may be of several kinds depending on the nature of the quality features. These are higher-the-better, lower-the-better and nominal-the-best [18]. The higher weld strength and lower weld width are desired in the present case, and hence the quality loss function is selected accordingly.

The quality loss function of the higher-the-better quality feature, l_i, can be expressed as

$$l_i = \left(\frac{1}{r}\sum_{r=1}^{n}\frac{1}{y_{ir}^2}\right) \tag{9.1}$$

where y_{ir} is the experimental result (of ith quality feature) for each experimental run and r is the number of repeating experiments for that experimental run.

The quality loss function of the lower-the-better quality feature can be expressed as

$$l_i = \left(\frac{1}{r}\sum_{r=1}^{n}y_{ir}^2\right) \tag{9.2}$$

To optimize more than one quality characteristic simultaneously, it is necessary to calculate the normalized quality loss. The normalization is done to bring uniformity by converting the quality characteristics in dimensionless quantity, to avoid different units and dimensions of different quality characteristics. The normalized values of quality loss can be calculated using [19]

$$l_{ij}^* = \frac{l_{ij}}{l_{i*}} \tag{9.3}$$

where l_{ij}^* is a normalized value of quality loss for the ith quality feature at the jth test setting, and l_{i*} is the highest quality loss for the ith quality feature among all the test settings. It is essential to remember that l_{ij}^* ranges from at least zero to at most one.

To calculate the total normalized quality loss (TNQL), L_j, must be assigned a weight factor for each quality characteristic considered in the optimization process, corresponding to each test condition. If w_i is the weight factor for the ith quality, k is the number of quality features and l_{ij}^* is a normalized quality loss connected with the ith quality at the jth test situation, then L_j can be calculated using [20]

$$L_j = \sum_{i=1}^{k} w_i l_{ij}^* \tag{9.4}$$

The next step is to calculate the multiple signal-to-noise ratio (MSNR) at each design point after the TNQL corresponding to each test condition is calculated. The MSNR, η_j, which corresponds to the jth test condition, is calculated as [21]

$$\eta_j = -10\log_{10}(L_j) \tag{9.5}$$

The objective is always to maximize the MSNR. The average value of all MSNR for a particular level of a process parameter is used to describe the effect of a process parameter at the same distinct level. The optimal level for a parameter is the level corresponding to the maximum average MSNR.

For each experimental run, the quality loss values of weld strength and weld width are calculated using eqs. (9.1) and (9.2) and those values are furnished in Table 9.5. For both the quality features in each experimental run, the normalized quality loss values are calculated using eq. (9.3) and are given in Table 9.5. The results of TNQL and MSNR calculated using eqs. (9.4) and (9.5) are presented in Table 9.5. The weighting ratio for both quality features is set as 1:1 in calculating TNQL, that is, each feature is of equivalent significance or comparative weighting.

Table 9.5: Results of Taguchi quality loss function analysis.

Exp.	Quality loss value (dB)		Normalized quality loss value		TNQL	MSNR (dB)
no.	Weld strength	Weld width	Weld strength	Weld width		
1	0.00068	7.45290	0.8095	0.5564	0.6829	1.6561
2	0.00055	7.12890	0.6548	0.5322	0.5935	2.2660
3	0.00058	7.02250	0.6905	0.5242	0.6074	2.1656
4	0.00084	7.12890	1.0000	0.5322	0.7661	1.1572
5	0.00039	9.30250	0.4643	0.6944	0.5794	2.3705
6	0.00061	6.86440	0.7262	0.5124	0.6193	2.0809
7	0.00043	9.67210	0.5119	0.7220	0.6170	2.0974
8	0.00058	7.23610	0.6905	0.5402	0.6153	2.1089
9	0.00027	11.90250	0.3214	0.8885	0.6050	2.1826
10	0.00025	11.35690	0.2976	0.8478	0.5727	2.4206
11	0.00054	6.20010	0.6429	0.4628	0.5529	2.5739
12	0.00056	6.30010	0.6667	0.4703	0.5685	2.4528
13	0.00018	13.39560	0.2143	1.0000	0.6071	2.1671
14	0.00025	9.36360	0.2976	0.6990	0.4983	3.0250
15	0.00036	7.45290	0.4286	0.5564	0.4925	3.0762
16	0.00053	5.33610	0.6310	0.3983	0.5146	2.8849

The MSNR is the overall quality index, which is the weighted combination of multiple quality features. Now, the objective is to maximize the MSNR [19]. The higher the value of MSNR, the better the multiple quality features. It is found that the experiment no. 15 has among the 16 experiments, the finest multiquality features in the current work, because it corresponds to the largest MSNR. Since the experimental design is orthogonal, the effect of each factors on the MSNR at each level can be extracted [20]. The MSNR responses are summarized and shown in Table 9.6 for each level of the welding parameters. Delta specified in Table 9.6 shows the variation of MSNR for different welding parameters. It is observed that for the selected parameter

Table 9.6: Response table for MSNR.

Welding parameters	Gray relational grade					
	Level 1	Level 2	Level 3	Level 4	Delta	Rank
Power, *A*	1.811	2.164	2.407	2.788	0.977	1
Welding speed, *B*	2.094	2.448	2.478	2.151	0.384	3
Stand-off distance, *C*	2.299	2.541	2.371	1.961	0.581	2
Clamp pressure, *D*	2.308	2.279	2.460	2.124	0.336	4

range, laser power causes the largest variation in MSNR. Thus, laser power is the most significant parameters for overall quality features, which is followed by stand-off distance, welding speed and clamp pressure.

The effect of factor levels on MSNR is shown in Figure 9.5. According to this graph, the optimum welding parameter setting for simultaneous optimization of weld strength and weld width is $A_4B_3C_2D_3$, that is, to maintain the laser power at fourth level, welding speed at 3rd level, stand-off distance at 2nd level and clamping pressure at third level. The ANOVA table for MSNR, furnished in Table 9.7, shows that all process parameters, except clamp pressure, are statistically significant.

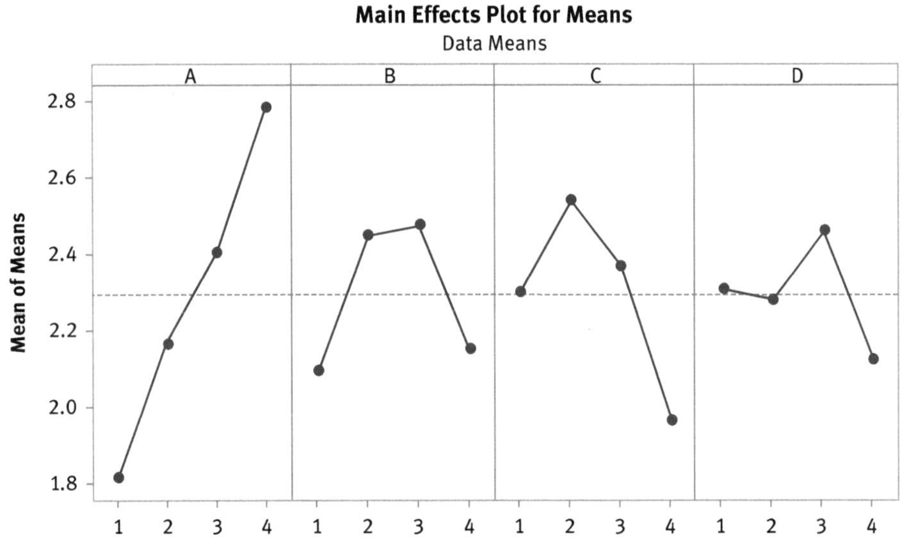

Figure 9.5: Main effects plot of welding parameters on MSNR.

After selecting the optimum levels of the welding parameters, the final stage is to check the enhancement of the quality features using the optimum parameter setting. A confirmation experiment is carried out by performing a test with optimum welding

Table 9.7: Analysis of variance for MSNR.

Sources	Sum of squares	Degrees of freedom	Mean squares	*F*-value	*p*-Value
A	2.02826	3	0.67609	47.57	0.005
B	0.47256	3	0.15752	11.08	0.039
C	0.71297	3	0.23766	16.72	0.022
D	0.22770	3	0.07590	5.34	0.101
Error	0.04264	3	0.01421		
Total	3.48412	15			

parameter setting, and the outcome is compared to previous best experimental outcome of experiment no. 15. The results of confirmation test are presented in Table 9.8. Improvement of the overall quality feature, MSNR, is found to be 0.1983 (6.45 %) at the optimum levels of parameter settings.

Table 9.8: Confirmation test results.

	Initial parameter setting	Optimal parameters
Level	$A_4B_3C_2D_4$	$A_4B_3C_2D_3$
Weld strength (N/mm)	52.67	54.19
Weld width (mm)	2.73	2.68
MSNR	3.0762	3.2745
Improvement of the MSNR = 0.1983		

9.5 Conclusion

Experimental investigation and parametric optimization of laser transmission welding of polycarbonate is reported in this chapter. Experimental plan is executed, and parametric investigation is carried out utilizing the Taguchi's methodology for four-level four-factor orthogonal array experiment scheme. Maximization of weld strength and minimization of weld width are considered as desired weld quality features. It is observed that the laser power has a major impact on total variation in weld strength, according to ANOVA results, and is followed by welding speed, stand-off distance and clamp pressure. From the ANOVA results, it is found that welding speed and stand-off distance have nearly the same effect, resulting in the source variation for weld width followed by laser power and clamp pressure. Taguchi quality loss function analysis is implemented to find out optimal parameter setting to optimize multiple quality features, simultaneously, by combining individual quality features into an overall quality index, that is, MSNR. It is observed that laser power produces the highest variation in MSNR for the chosen parameter range. Laser power is therefore the most important parameters for overall weld quality, followed by stand-off distance, welding speed and clamping pressure. The optimum welding parameters for optimizing weld strength and weld width, simultaneously, determined by the Taguchi quality loss function analysis is $A_4B_3C_2D_3$. The results show that in laser transmission welding, the application of the Taguchi quality loss function evaluation results in an effective improvement of the weld quality.

References

[1] Stokes, V.K. Thermoplastics as engineering materials: The mechanics, materials, design, processing link. J Eng Mater Technol 1995, 117, 448–455.

[2] Jansson, A., Kouvo, S., Salminen, A., and Kujanpää, V. (2003) 'The effect of parameters on laser transmission welding of polymers', Proceedings of the 22nd International Congress on Applications of Lasers & Electro-Optics, Jacksonville, Florida, USA, Vol.95, LMP Section A, pp.124–133.

[3] Acherjee, B., Kuar, A.S., Mitra, S., and Misra, D. Empirical modeling and multi-response optimization of laser transmission welding of polycarbonate to ABS. Lasers Manuf Mater Process 2015, 2(3), 103–123.

[4] Acherjee, B., Kuar, A.S., Mitra, S., and Misra, D. Laser transmission welding of thermoplastics: An overview of experimental findings – process, development and applications. J Manuf Technol Res 2011, 3(3/4), 211–236.

[5] Bonten, C., and Tüchert, C. Welding of plastics-Introduction into heating by radiation. J Reinf Plast Compos 2002, 21(8), 699–710.

[6] Acherjee, B., Kuar, A.S., Mitra, S., and Misra, D. Laser transmission welding of polycarbonates: Experiments, modeling, and sensitivity analysis. Int J Adv Manuf Technol 2015, 78(5–8), 853–861.

[7] Bachmann, F.G., and Russek, U.A. (2002) 'Laser welding of polymers using high power diode lasers', Proceedings of Photonics West Conference, San Jose, CA, USA; Proc. SPIE, Vol.4637B.

[8] Baylis, B. (2002) 'Welding thermoplastic elastomers to polypropylene with a diode laser', Proceedings of the 21st International Congress on Applications of Lasers & Electro-Optics, Scottsdale, Arizona, USA, Vol.94, LMP Section E.\

[9] Kagan, V.A., Bray, R.G., and Kuhn, W.P. Laser transmission welding of semi-crystalline thermoplastics: Part I: Optical characterization of nylon based plastics. J Reinf Plast Compos 2002, 21(12), 1101–1122.

[10] Kagan, V.A., and Pinho, G.P. Laser transmission welding of semicrystalline thermoplastics – Part II: Analysis of mechanical performance of welded nylon. J Reinf Plast Compos 2004, 23(1), 95–107.

[11] Grewell, D., Rooney, P., and Kagan, V.A. Relationship between optical properties and optimized processing parameters for through-transmission laser welding of thermoplastics. Reinf Plast Compos 2004, 23(3), 239–247.

[12] Kagan, V.A., Bray, R., and Chambers, A. Forward to better understanding of optical characterization and development of colored polyamides for the infra-red/laser welding: Part I – Efficiency of polyamides for infra-red welding. J Reinf Plast Compos 2003, 22(6), 533–547.

[13] Kagan, V.A., Chambers, A., and Bray, R. Forward to better understanding of optical characterization and development of colored polyamides for the infra-red/laser welding, part II – Family of colored polyamides. J Reinf Plast Compos 2003, 22(7), 593–603.
[14] Jansson, A., Kouvo, S., and Kujanpää, V. (2004) 'Preliminary investigations of laser welding of plastics in mass production', Proceedings of the 23rd International Congress on Applications of Lasers and Electro-Optics, San Francisco, California, USA, Vol.97, paper504.
[15] Acherjee, B., Kuar, A.S., Mitra, S., and Misra, D. Effect of carbon black on temperature field and weld profile during laser transmission welding of polymers: A FEM study. Opt Laser Technol 2012, 44(3), 514–521.
[16] Prabhakaran, R., Kontopoulou, M., Zak, G., Bates, P.J., and Baylis, B.K. Contour laser – Laser-transmission welding of glass reinforced nylon 6. J Thermoplast Compos Mater 2006, 19, 427–439.
[17] Acherjee, B., Kuar, A.S., Mitra, S., and Misra, D. A sequentially integrated multi-criteria optimization approach applied to laser transmission weld quality enhancement – A case study. Int J Adv Manuf Technol 2013, 65(5–8), 641–650.
[18] Phadke, M.S. Quality Engineering using Robust Design, 1989, Prentice Hall, Englewood Cliffs, New Jersey.
[19] Aslan, N. Multi-objective optimization of some process parameters of a multi-gravity separator for chromite concentration. Sep Purif Technol 2008, 64, 237–241.
[20] Acherjee, B., Kuar, A.S., Mitra, S., and Misra, D. The use of Taguchi quality loss function to optimize multiple quality characteristics in laser transmission welding process of thermoplastics. ASME Early Career Tech J 2009, 8(1), 208–213.
[21] Antony, J. Simultaneous optimization of multiple quality characteristics in manufacturing processes using Taguchi's quality loss function. Int J Adv Manuf Technol 2001, 17, 134–138.

Section V: **Environmental Issues**

Emmanuel Baffour-Awuah, Stephen Akinlabi, Tien-Chien Jen

10 The polyethylene pollution challenge: a review

Abstract: The benefits of polyethylene (PE) to both developed and developing nations have been practically and literally espoused by various scholarships and in numerous literatures and therefore cannot be underestimated. However, the use of polythene has unwarranted consequences on the environment. Its direct and indirect negative consequences on biological organisms also need much to be desired. Though various methods are used to deal with the menace, challenges still prevail. The paper first dilated on the structure and properties of PP, which provides its inherent characteristics. The aim of the chapter is to encourage further empirical and theoretical studies in the area of PE research. The objective of the chapter is, however, to highlight direct and indirect environmental and health costs of plastics in general and PE usage in particular. Flooding, water, air, land and visual pollution were identified as some of the effects of PE usage. With regard to public health effects, diseases such as infertility; cardiovascular, nervous and reproductive diseases; sexual immaturation; aggressive behavior, breast cancer and animal hermaphroditism arising from the presence of endocrine-disruptor compounds such as bisphenol A and bisphenol S were also identified. Others include insulin resistance and increased waist circumference, which emanates from ingestion of di-(2-ethylhexyl phthalate). The second objective is to argue that there are opportunities for further research into PP with regard to knowledge enhancement toward PE-natural fiber composite formation and biodegradation as a means of managing the environment. Future researchers, practitioners and industrialists could also concentrate on empirical simulation research toward actual public health consequences of PP on humans. It is expected that research in these areas could help save nations billions of dollars through innovations as well as inexpensive techniques, procedures and processes.

Keywords: Environmental pollution, plastics, polyethylene, plastic-biofiber composites

Emmanuel Baffour-Awuah, Department of Mechanical Engineering Science, Faculty of Engineering and the Built Environment, University of Johannesburg, Auckland Park, South Africa; Department of Mechanical Engineering, School of Engineering, Cape Coast Technical University, Cape Coast, Ghana

Stephen Akinlabi, Department of Mechanical & Industrial Engineering Technology, Faculty of Engineering and the Built Environment, University of Johannesburg, Auckland Park, South Africa; Department of Mechanical Engineering, Covenant University, Ota, Nigeria

Tien-Chien Jen, Department of Mechanical Engineering Science, Faculty of Engineering and the Built Environment, University of Johannesburg, Auckland Park, South Africa

https://doi.org/10.1515/9783110655049-010

10.1 Introduction

Plastic as an engineering material has been quite beneficial to both developed and developing countries since the first fully synthetic plastic, bakelite, was invented in New York in 1907 by Leo Bakeland [1, 2]. In spite of the manifold benefits of plastics is accompanied with some challenges, particularly within the public health and environmental sectors [1]. Various types of plastics are in use at present. Common among them are polypropylene, polyvinyl chloride, polyamides and polyethylene (PE). Though these plastics are widely used, PE appears to be the most commonly used plastic, accounting for about a third of all plastics used in the manufacturing and consumption sectors globally (Polymer Converters Limited, 2006).

It is estimated that in the United States alone, petroleum fuel is used to manufacture plastic water bottles that could energize a million vehicles; and about 99% of all plastics are made from petroleum products. Considering that, over a third of plastic consumption is PE, and that US$40 billion of linear low-density polyethylene (LLDPE), US$33 billion of low-density polyethylene (LDPE) and 30 million tons of high-density polyethylene (HDPE) are marketed globally, recycling of PE would be a better option to reduce plastic menace to society. According to a report, global PE market was expected to reach around USD163 billion in 2017 and is expected to reach approximately US $215 billion by 2024, at a compound annual growth rate ("a business and investing specific term for the geometric progression ratio that provides a constant rate of return over the time period") of a little above 4% between 2017 and 2024 [3–5].

PE has been found most useful as a result of its characteristics, which include cost-effectiveness, versatility, imperviousness to water and less energy requirement in its production, among others. Unfortunately, these innumerable advantages are accompanied with challenges to both living organisms and the environment. PE may be transparent, soft, flexible, moldable and/or biodegradable, enabling its applications as a suitable innovative engineering material in a wide range of inexpensive uses such as carbonated drink bottles, plastic films, peanut butter jars, microwave packaging; shower curtains, clamshell packaging, floor tiles, siding, outdoor furniture; milk jugs, molded plastic cases, detergent bottles, plastic bottles; and water container bags.

Transparent polythene bags are used to package water. This packaged water is usually referred to as "pure water" or "sachet water." In developing countries such as Nigeria, India and Ghana, sachet water is considered as the modern low-cost technologically innovative way of rendering water [6]. This has contributed to waste sachets to become an environmental challenge, which demands critical attention. Its risk on environment and public health cannot therefore be undoubted. Inasmuch as plastic applications continue to throw challenges to humanity, its use cannot be curtailed due to the numerous benefits. In order to ensure sustainable use of plastics in general and PE in particular, innovation should be encouraged to reduce, if not eliminate, the challenges associated with the production and consumption of the material.

One way by which these challenges could be minimized is the introduction of plastic-natural fiber composites, a recycling technique that makes use of waste plastic and natural fiber to produce a new material that is relatively better degradable than homogeneous plastics. Adopting the polythene-natural fiber composite technique could therefore be one of the best options in dealing with the waste polythene challenge in developing countries.

10.2 Origin of polyethylene ($[-CH_2-CH_2-]_n$)

PE, polythene or poly (meth $(C_2H_4)_n$ ylene) with chemical formula $[-CH_2-CH_2-]_n$ and the most widely used common plastic, was first formed accidentally in 1897 as a precipitate of white flakes from diazomethane solutions [7, 8]. However, due to its scarcity at the time, no further investigations were made. The major ingredient of PE is petroleum or natural gas, though ethylene could also be obtained from renewable materials such as plant matter.

Industrially, PE was first synthesized in 1933 by Eric Fawcett Imperial and Reginald Gibson; also accidentally at the premises of Chemical Industries (ICI) in Northwich, England. Reproduction of the material was difficult to reproduce until 1935 when a colleague chemist in ICI, Michael Perrin, developed the accidental procedure into a high-pressure synthesis of PE. This procedure became the industrial procedure for the manufacture of LDPE from 1939 [8]. Commercial production of the plastic (large scale), under the license of ICI commenced in 1944, using the ICI process by Bakelite Corporation at Sabine Texas and Du Pont at Charleston West Virginia, both in the United States.

Without the development of catalysts to facilitate the production of PE, commercial manufacture would have been impossible. Catalysts ensure that reduction is facilitated at mild temperatures and pressures. In 1951, Robert Bauks and J. Paul Hogan at Phillips discovered a chromium-trioxide-based catalyst to facilitate the polymerization process [9]. By late 1953, a German Chemist Karl Ziegler had discovered the "Ziegler catalyst," a catalyst based on titanium halides and organo-aluminum compounds, which performed at milder pressures and temperatures. The Ziegler technique required more expensive engineering though temperatures and pressures of manufacture were lower. The relative advantages of both techniques encouraged their applications in commercial production of PE by the end of 1950s. HDPE was the product of the Phillips and Ziegler catalyst technique. By the close of 1970s, other catalyst-based techniques were developed. For example, the Ziegler method was improved by the addition of magnesium chloride catalyst; metallocenes as soluble catalysts were also reported as effective in the production of plastics in 1976 by Walter Kaminsky and Hansjorg Sinn. Application of the Ziegler and metallocene-based catalyst groups is very versatile in the copolymerization of ethylene, and therefore form the foundation of various types of PE such as LDPE, LLDPE, very LDPE and ultra-high molecular weight polyethylene (UHMWPE) [8].

10.3 Properties of polyethylene

The characteristics of PE such as mechanical, thermal, chemical, electrical and optical are depicted in its various properties as a plastic material. In terms of mechanical properties, PE is hard, low in strength, rigid and has low friction. It has high impact strength and ductility. Under persistent force, PE exhibits high creep strength though this characteristic may be lowered by adding short fibers. When touched, PE feels waxy [10].

Melting point of PE varies according to the type. The four main forms of PE include LDPE, HDPE, LLDPE and medium density polyethylene (MDPE). The low melting point of PE enhances its usefulness as a plastic material. While the melting point of HDPE is about 80 °C, the melting point of common-grade MDPE and HDPE ranges between 120 and 180 °C (248–356 °F). The melting point of average and commercial LDPE is between 105 and 115 °C (221–239 °F).

Chemically, PE is partially crystalline in general due to its symmetric molecular structure. Crystallinity increases with density, mechanical and chemical stability. Thus, the macromolecules of PE are not bonded covalently. PE is made of saturated nonpolar and high-molecular-weight hydrocarbons. Most PE has good chemical resistance to strong acids, strong bases, gentle oxidants and reducing agents. At elevated temperatures they may dissolve in chlorinated solvents such as dichlorobenzene and trichloroethane; and aromatic hydrocarbons such as xylene and toluene with the exception of the cross-linked types. At room temperature, crystalline samples of PE will not dissolve in these solvents [11]. When ignited, PE burns slowly with a flame characterized by blue, with yellow tip and a paraffin-scented odor (candle flame type). Upon separation from ignition source it continuously burns and drips off. When exposed to sunlight PE becomes brittle. Carbon black of PE may be used as an ultraviolet stabilizer (application of burn test). Though water vapor and polar gases are lower in permeability, carbon dioxide, oxygen and flavorings can pass through PE very readily.

Electrically, PE is a good insulator; offering elevated tracking resistance thus sustaining surface discharges without structural or electrical destruction at currents below 10 ma in the presence of dust and moisture. Though it can be easily charged electrostatically, this defect could be reduced by adding antistatic agents, graphite or carbon black. Chung [12] records that PE may be optically transparent (almost clear), milky translucent (milky-opaque) or opaque depending on the thickness of film and thermal history. PE with larger wavelength than the wavelength of visible light has more opaque crystals. Thus LDPE is the greatest, LLDPE slightly less and HDPE the least transparent.

10.4 Types of polyethylene

With reference to volumes of PE sold globally, the prominent among various grades are LDPE, LLDPE, MDPE and HDPE. Other grades include UHMWPE, ultra-low-molecular-weight PE, high-molecular-weight PE, high-density cross-linked PE and chlorinated PE. PE is classified according to the branching and density. For this reason, the mechanical properties of the plastic depend significantly on factors such as molecular weight, type of branching and crystal structure. For the purpose of convenience and objectives of this chapter, the grades prominently sold globally shall be discussed, which include HDPE, LLDPE, LDPE and MDPE.

HDPE has a low branching grade with density greater or equal to 0.941 g/cm^3. Catalysts used in its production are Ziegler–Natta catalysts, metallocene catalysts or chromium/silica catalysts. HDPE has high tensile strength, thus encouraging its application in the production of detergent bottles, garbage containers, milk jugs, butter tubs and water pipes, among others. About one-third of toy manufacturers employ HDPE. By 2007, the global HDPE demand market had reached over 30 million tons [13].

MDPE is characterized by a density range of 0.926–0.940 g/cm^3, and produced by the Ziegler–Natta, metallocene or chromium/silica catalyst techniques. Comparatively, shock, drop and stress cracking properties are better than HDPE. However, notch-sensitive property is less than HDPE. It is therefore more suitable in the production of shrink films, packaging film, carrier bags, sacks; gas pipes and fittings; and screw closures.

LDPE is characterized by a density range between 0.910 and 0.940 g/cm^3 with a high grade of short- and long-chain molecular branching. It has relatively lower tensile strength and higher ductility. The radical polymerization process employed does not require catalyst. Molten LDPE has a unique yet desirable flow property. These characteristics inure to its benefit as a material for production of rigid containers and plastic film applications such as film wrappers and plastic bags and sachets used to package sachet water.

The density of LLDPE ranges between 0.915 and 0.9925 g/cm^3, with relatively high tensile strength and high impacts and punctured resistance than LDPE. When thickness is comparatively low, LDPE films can be inflated and has better environmental stress-cracking than LDPE, though processing is difficult. Like LDPE, its applications include film bags, sheets, and water sachets; cable coverings, lids, containers, toys, buckets and pipes; agricultural films, bubble wrappers, multilayer and composite films. These applications are possible due to its relative transparency, toughness and flexibility. By 2013, the annual market volume had hit US$40 billion per annum. In its application for film bags, sheets and sachets, LLDPE may be a good substitute for LDPE [4].

10.5 The polyethylene challenge

Environmental challenge is one of two basic challenges that plastic applications throw as a concern to modern societies. The challenges of plastic applications are basically categorized into two folds: environmental and public health concerns [1]. In 2015 alone, overall environmental cost of plastic applications with reference to consumer goods sector was about US$139 billion [14]. The use of polythene contributes about a third to this cost [3]. Global concerns are being raised [15]. In Ghana, the use of PE bags and bottles has become so commonplace that concerns are also being raised about its use. Nevertheless, plastic bags and bottles provide relatively better satisfaction both at home and market. It is a convenient way, painless and easy means of conveying market items to homes [16, 17].

However, the effect on the environment cannot be underestimated. The impact of plastic use on the environment is manifold. First, the origin of plastics creates environment consequences [17]. PE is basically produced from crude oil and natural gas. During manufacturing, a substantial amount of pollutants are released into the atmosphere, lithosphere as well as the hydrosphere. As petroleum-based product, pollutants such as nitrogen oxides, sulfur oxides, carbon dioxide and heavy metals such as antimony, cadmium and mercury are released into the environment. In the United States, 12 million barrels of oil is exposed in manufacturing and 100 billion PE bags annually [18]. Source of energy employed during manufacturing, if petroleum-based, further adds to the quantum of pollution meted out to the environment. Studies show that between 60 and 100 million barrels of oil are needed to produce annual plastic bags globally [17]. Globally, between 500 and 1 trillion plastic bags are injected into the environment annually [19]. The Independent reports that over 50 million garbage bags are deposited into the Pacific Ocean near Australia annually, forming “plastic-soup” twice the size of the continental United States; and in Britain, it is about 80% of this size [17].

Second, the manufacturing challenge to the environment, that is, when PE has been manufactured, used and injected into the environment, has been found to be relatively durable considering the fact that PE, for example, is relatively durable considering the fact that it requires approximately 400 years or more to biodegrade [17]; and 1,000 years to completely decay [18, 20]. Thus PE accumulates in the course of polluting the environment. In addition to this, PE bags are transported from the manufacturer to retail points. It is of interest to note that about 25% of PE bags used in western countries (the United States, Canada and Europe) are produced in Asia [15]. The source of fossil fuel used to transport the commodity is thus a source of pollution in terms of production and transportation. In 2014, annual consumption of crude oil to manufacture plastic bottles alone was estimated to be about 17 million barrels [15, 19].

Third, PE bags and bottles create environmental menace in and around landfills. The Clean Air Council in the United States reports that, in the United States alone, residents dispose 2–5 million plastic bottles every hour, amounting to about 60 million plastic bottles per day [19]. According to the Sierra Club, 24 billion plastic bottles is considered as trash in the United States on annually basis. Plastic bags and bottles that end up in landfills as trash may not decompose any time soon. Plastic bags and bottles are usually produced from PE. They are, therefore, a threat to wild life and natural areas and reduce aesthetic nature of the environment. The cost of cleaning up litter and trash into landfills may not therefore be much to desire. According to the Container Recycling Institute [21], the tax payer spends millions of dollars each year to deal with trash and litter in making the environment clean. Even in landfills, PE bags may escape by virtue of wind blowing, leaving them entangled on nearby vegetation [15]. Though it is considered that landfills are created with intention to serve a particular locality, population and for a period of time, many a landfills have suffered from overloading, undue and unexpected pressure as a result of two reasons. First, there is too much waste burden of the facility; and second, inability to locate a new site as a landfill. This situation is currently occurring in Accra and Kumasi, both in Ghana, where metropolitan mayors have recently ordered discarded landfill site to be opened for reuse since the metropolis is engulfed with garbage, and large proportion of which is PE bags and bottles.

Fourth, burning PE as means of disposing them off has its fair share of environmental pollution. Smoldering and incinerating plastics release toxic fumes to pollute the atmosphere. Gases such as carbon dioxide, a greenhouse gas, are injected into the atmosphere. The effect could be global warming with consequential flooding and stormy weather conditions. Double-bagging of plastic bags due to overloading, arising from tear or breakage of single bag handles, adds to the quantum of waste PE materials that are incinerated [18, 22]. Chemicals so released from burning of plastic bags may also pose health risk to living organisms and may donate to break down of the ozone layer [23].

Lastly, waste plastics such as PE bags and bottles may create flooding, particularly, within highly populated communities, and eventually ending up in the ocean. It is estimated that between 60% and 80% of all marine debris is made up of plastics; and in some locations between 90% and 95%. If improperly managed, plastic waste may find its way into the oceans, accumulating if single used in its application, within a related short period of time. As principal support system of marine ecosystem, the prevalence of plankton and algae is threatened by the presence of floating plastic debris as they tend to prevent sunlight from reaching the bottom of the ocean, thus breaking the universal food chain system. Plastic debris in the ocean has the capacity to also entangle and choke ship propellers. Out of a total of 20 billion pounds of plastics produced annually, 20% end up in the ocean; while 30% of that enters the ocean sinks, the remaining floats, forming plastic islands. Fifty percent goes to landfills while the remaining litter in the environment [14, 15, 24, 25].

According to the Sierra Club, plastic bags do not break down completely in water but rather decomposes into plastic debris, and neutrally into plastic dust (with the release of the biotoxin polychlorinated biphenyls) [16]. These chemicals could be ingested by humans through the food chain, posing as health threat through direct and indirect consumption of marine plants and animals [26]. Wisconsin Department of Natural Resources reports that waste PE bags when indiscriminately littered and disposed could lead to clogged sewer gates and gutters, resulting in flooding [16]. They may eventually end up in oceans as thrash. It has been confirmed by the National Oceanic and Atmospheric Administration's National Ocean Service Marine Debris Program that there exists a garbage patch in Great Pacific consisting of a thrash mainly composed of debris of plastic waste such as plastic bags as they decompose (Environment Natural Environment, 2010). The challenge is that these pellets of debris and the associated toxins may be ingested by ocean organisms [26].

A 2009 United Nations Environment Program report also indicates that plastic bags from most of the debris within the Mediterranean and the Red sea contribute to the destruction of cosmetics of some waterways and pristine beaches across the globe. It is estimated by the Californians Against Waste and the Marichkville Council of Australia that 100,000 marine turtles, whales and birds die every year as a result of plastic deposition in the ocean due to suffocation and/or toxification. Ingestion of plastic debris by marine organisms may eventually lead to blockage of digestive tract and stomach or intestinal obstruction, which may result in starvation and malnutrition, reduced reproductive capacity, degeneration of general quality of life, emaciation and loss of life. It is estimated that the consumer goods sector cost of plastic marine debris is $4.7 billion per annum [14, 15, 18, 27].

10.6 Public health challenge

The perpetual contact between human body and the ingredients of plastics due to incessant use of the material has brought in its wake some public health challenges to the global community [1]. Studies have shown that toxic plastic ingredients rarely accumulate in the body except accidental exposure, which results in instantaneous buildup in the gastrointestinal tract [28]. Thus, there is a continual balance between steady-state exposure, exposed metabolism and excretion of these toxins in the human body. Since plastic materials are used from cradle to grave of individual lifetime, plastic ingredients are ingested throughout individual's lifetime. Determining a control group for experimentation of the effects of plastic use on the human body has therefore been extremely difficult. Ethical considerations have also aggravated the situation due to toxicity of these ingredients on the human body. For these two fundamental reasons, most studies related to the effects of plastic use by human societies have largely been carried out on controlled animal populations as experimental subjects to replace human populations.

Numerous compounds that have negative effects on human health have been identified to be present in plastics. Most popular among are di-(2-ethylhexyl phthalate) (DEHP), bisphenol A (BPA), bisphenol S (BPS) and perfloro-octanesulfonate (PFOS). For example, a study in the United States identified that 95% of adults have BPA in their urine. Exposure to these compounds has been found to be detrimental to animal health through various experimentations. Particular negative health effects include male infertility; cardiovascular, nervous and reproductive diseases; sexual immaturation and aggressive behavior [29–34]. The potential harm of BPA was first identified by the US Food and Drug Administration (FDA). BPA as an artificial compound is characterized as an endocrine disruptor (ED). It functions in animals, including humans, to mimic the role of natural hormones in the human body [35]. Though there is little evidence to prove its effects on humans, the University of Cincinnati [35] study identified that the exposure of BPA does not depend on the age of plastic but rather on temperature at which the plastic was exposed. The study further noted that the rate of release of BPA into the environment may be 15–55 times higher at increased temperatures up to boiling point of water. Thus, when plastics containing BPA are heated in water, the quantum of compound released increases with respect to temperature of water.

BPA is found in certain plastics used in the manufacture of consumer products such as water and food storage containers. A study conducted by Manshack et al. [36] tried to identify the genetic pathway that may be altered in the brain due to BPA exposure during fetal development, relying on earlier studies that BPA could alter sexual function and behavior in turtles. The study showed that BPA exposure can lead to disruption of genetic pathways confirming that ingestion of BPA could negatively influence reproductive systems of animals. The effect of BPA on hyperactivity has also been confirmed in a study by Kinch et al. [37]. Using zebrafish as subjects, with application of a minute concentration of the compound, there was significant change in the time of brain neuron formation at the embryonic stage during reproduction. "These findings are important because they support that the prenatal period is a particularly sensitive stage," confirming how exposure to BPA may alter brain development [37]. For the first time, a study has also shown that there is a direct effect of low-dose BPA exposure on prostate disease and pathology [38]. The study identified that the endocrine-disrupting compound BPA can "reprogram the developing prostrate, making the gland more susceptible to precancerous lesions and other diseases later in a man's life," by mimicking the hormone estrogen.

Since the realization of BPA as a potential harm to human health, the FDA has imposed a ban from using it in the manufacture of various consumer products such as infant bottles and spill-proof cups for toddler use [39, 40]. Similarly, in Canada and the European Union, BPA has been prohibited in plastics used in producing baby bottles (European Food Safety Authority). The response to these bans has culminated in the replacement by BPS as a "BPA-free" substitute. It was believed that BPS was not an EDs. Recent studies have, however, proven that BPS is also an ED, which is able to mimic hormone activities. For example, Kinch et al. [37] using zebra fish as subjects observed that even minute quantities of BPS can alter brain development substantially. The study identified that BPS introduced during embryonic development could lead to neuron-connection-circuit disconnections and misconnections by increasing neurons born too early by twice and delaying neurons born lately by twice. For unexposed fish, neurons that were born increased by 180%. These neuron developments, the study concluded, led to hyperactivity.

DEHP is a phthalate. Phthalates are plasticizers, that is, chemicals used in the manufacture of plastics to increase transparency, flexibility, durability and longevity. They have been identified as EDs. DEHP has been found to leach readily since it is unable to bond with plastic materials [41]. Various studies relating rodents and humans have identified correlations between DEHP exposure and negative public health effects such as male and female reproductive systems; insulin resistance and increased waist circumference [41, 42]; liver and testes damage; metabolic interference, hepatocarcinogenic effects and endocannabinoid system disruption [43].

PFOS, a polyhalogenated flame retardant [and a surfactant (a substance when dissolved in a liquid tends to reduce its surface tension)] used in plastic manufacture has been studied as to whether it is potentially harmful to human health [1]. Romano et al. [29] report that compounds found in the environment which are able to mimic the hormone estrogen has potential harmful effects since they influence the gonad functions; and that ED compounds such as PFOS could also affect cardiac functions. Thus, PFOS mimic estrogens and may, therefore, have harmful effect on heart functions. The report also indicates that these chemicals could also have deleterious effects on nervous and reproductive systems.

As plasticizer, phthalates are employed in the manufacture of various types of plastics, including PE. Studies have shown that phthalates can have deleterious consequences on human and animal health. As indicated earlier, the harmful effects of phthalates include hepatocarcinogenic effects; metabolic interference; liver, kidney and testes damage; cardiovascular effects; and reproductive consequences. Other harmful effects on human and animal health include neurotoxicity, hyperactivity, waist circumference (obesity), insulin resistance (diabetes), heart physiology and blood clothing carcinogenic effects. It is estimated that about 70% global productions of plasticizers are phthalates [43]. What makes the application of phthalates in the production of plastics more frightening is that, aside the potential and established harmful effects, there are doubts about the safety thresholds for these chemicals. For example, Habert et al. [44] studied that employing humans and rats (testes) revealed different responses in about 66% of all cases using BPA, one of phthalate group. The study therefore recommended a re-evaluation of the daily acceptable intake for EDs used in the manufacture of plastics.

An interesting finding was also made by Ayamba [45] in a study entitled "Assessment of phthalates migration in PE Food packages: A Case study in Ghana."

Five different types of PE packaging materials were examined as to whether they contain one kind or the other. Materials examined include black PE bags, thick plain PE bags/films (takeaway bags as popularly referred), PE plastic bottles and PE food containers. The phthalates investigated include dibutylphthalate, DEHP, diethyl phthalate and benzyl butyl phthalate, which are all EDs and may therefore lead to one or more of the public health challenges indicated earlier on in this chapter. The results of the study indicated that all the four phthalates were present in the PE packaging materials. It is evidential that phthalates are used in the manufacturing of polythene for food packaging and that they find their way into both food and the environment [46, 47]. Similar evidence exists for plastic drinking containers made of PE [48–50].

It is evidential that bottles made of PE, particularly polyethylene terephthalate, can leach antimony, a toxic metal used in the manufacture of the bottles; and that bottled water may have concentrations of antimony 100 times that in clean underground water, about 2 parts per trillion [51, 52]. It is also evident that the longer the bottles stay on shelves the greater antimony leach. The study further showed that leaching increases as bottles are exposed to sunlight. Furthermore, the study observed that antimony concentration increases with increase in temperature. Brominated compounds have also been observed to leach into these bottles [53, 54].

It is well noted that both antimony and brominated compounds are toxic to humans and animals. While antimony is an ED, brominated compounds displace iodine in the body due to the presence of bromine in the compound. Through accumulation, brominated compounds can therefore initiate paranoia and other psychotic symptoms: attacking the central nervous system and acting as a depressant. With the potential and established harm on human and animal health and the environment authenticated; and with reference to phthalates in sachet water PE, there is the need to develop means to curtail the use of plastics in general and PE in particular within modern societies; or reduce its abundance in the environment. It appears the former is impossible looking at the various uses, benefits and permeation of plastic use worldwide. It appears that the latter is more practicable if the use of plastics, specifically PE, is to be used without compromising the quality of life of human societies now and the future.

10.7 The way forward

The plastic age and subsequent mass production of consumer plastic products has brought in its wake negative environmental and public health challenges, which were oblivious to researchers, manufacturers and consumers at the beginning of the period. The structure and resulting properties of PE has inured to it the discredit of creating both public health and environmental challenges. While it may be difficult in dealing with the direct public health challenges and indirect health hazards caused by environmental disposal of the plastic, the direct environmental effects could be reduced to the minimum. With regard to direct public health challenges, further studies need to be conducted in confirming the precise health effects of PE use on human beings and other organic matter. Environmental pollution of PE breeds negative consequences, including indirect public health challenges through the environment and direct ones such as eating and drinking from PE packages. Indirect pollution effects include those through the food chain. Direct environmental consequences such as flooding, visual pollution and air pollution could be dealt with through application of technology. With these challenges in the face of the current generation it is high time technologies are innovated to deal with the naked canker. Adopting effective and efficient waste PE disposal techniques could go a long way to reduce direct PE pollution. The techniques available in dealing with the situation include landfilling, waste incineration, biodegradation and recycling.

Comparatively, the use of plastics, including PE with other materials of similar use such as glass, paper and metal in various applications, has some advantages. For example, replacing plastics in consumer products and packaging with these substitutes to perform similar functions could increase environmental cost. In terms of weight, the ratio of plastics to alternatives is about 1:4. Thus in the entire consumer goods sector, the 84 Mt of plastic used in the consumer products and packaging in 2015 would have been substituted by over 342 Mt of alternative material. Similarly, the amount of fossil fuel used in producing these alternative materials is far higher than that for producing plastics. In the United States, 4.6% of petroleum is consumed, 2 million barrels as fuel and 329 million barrels as feedstock in the manufacture of plastic materials [14].

In spite of these benefits landfilling is bereft with several challenges. First, landfilling causes almost 100% sequestration of carbon in plastics, thus permanently burning both material matter and energy during the process [55]. In relative terms with alternative materials, such as glass and metal, some energy and material matter could be retrieved. Second, the use of protective liners forming a protective layer between the soil and landfills to prevent leachates from entering the soil in state-of-the-art engineered landfills may tear, burst, fracture or tear, allowing the soil facility to be contaminated with leachates. The long-term health risk in this regard cannot be underestimated [56], arising from toxic plastic components and other landfill pollutants. In developing countries where there is virtually no engineered landfill sites, and that garbage and refuse are thrown into seemingly any suitable landfill site, the degree of contamination and pollution can be quite colossal. Finally, it is well appreciating that land space is not infinite. For instance, in countries such as Japan and Denmark, scarcity of land is beginning to rage its ugly head, driving the economy to rely much on garbage incineration as a land conservation technique. In other developed countries, time will tell as to whether land is in abundance supply or not. In many populated areas in both developed and developing countries, the unavailability of landfill sites appear to becoming increasingly scarce than ever as cities metropolis, municipalities and urban communities continue to expand with population explosion in such areas. The rural–urban drift; developing country–developed country migration; as well as war-zone and non-war-zone migration continue to aggravate waste disposal challenge such as landfilling in these areas.

Incineration, an act of burning or reducing substances to ashes, may be employed as a waste disposal technique. This type of incineration is available in terms of volume of waste disposed: small scale would normally include waste of 2–40 households; and industrial-sized municipal waste incineration. Waste may also be burnt openly or in engineered facilities. For example, in Ghana, plastics and other wastes are usually burnt in the open except in few formal organizations that do so in designed facilities. In some parts of the developing world and rural America, waste may be burned in barrels rather than landfills. Generally, most landfill waste is burned in the open in many parts of the developing world. This medium of waste disposal has some benefits. For example, it reduces material waste mass and volume significantly: recovers some portion of energy used to produce the waste material; and lead to disease prevention. Nevertheless, incineration does not necessarily lead to sustainability nor is it free from imposing health risk to human, plants and animals. For instance, incineration can inject toxins such as polychlorinated dibenzo-*p*-dioxins/polychlorinated dibenzofurans into the atmosphere. Ash from incinerated plastics and other solid waste materials may also contain heavy metals that are toxic to human health. These ashes and heavy metals may leach into soil and subsequently find its way into groundwater resources underlying the waste deposition sites [57], posing health risks to living organisms.

Biodegradation, a waste disposal technique, may be defined as "degradation that results from the action of naturally occurring microorganisms such as bacteria, fungi and algae." Biodegradable plastic technology may also be used to dispose plastics. Thus a biodegradable plastic could disintegrate into smaller and smaller microparticles, though its harmful effects on the environment may still persist. Currently, biodegradable plastics are being developed, such as polylactic acid and polyhydroxylalkanoate, though they do not meet the ideal disposable plastic criteria. Compostable plastics are those that "undergo degradation by biological processes during composting to yield carbon dioxide, water, inorganic compounds and biomass at a rate consistent with other known compostable materials and leaves no visible, distinguishable or toxic residue" [58]. Bioplastics are derived from edible products such as maize and molasses, thus competing with human and animal food supply [59]. With the current price levels of crude oil, biodegradable plastics may be more expensive than conventional plastics, though the environmental and public health benefits may far outweigh the costs.

Nevertheless, advances have been made with regard to biodegradation of plastics that are traditionally considered to be nonbiodegradable. A major challenge associated with PE is its inability to biodegrade, hence accumulating, without distinct treatment. After the Fukushima disaster in 2011, the issue of disposing plastics in an environmentally friendly manner became a topical environmental challenge in Japan though a definite and aggressive effort began in 2008 [9]. Modern advances to deal with the issue include the discovery of *Pseudomonas fluorescens*, with the assistance of *Sphingomonas*, to degrade over 40% weight of plastic bags in less than 3 months [60]; the isolation of a thermophilic bacterium *Brevibacillus borstelensis* (Strain 707) from a soil sample, which feeds on PE as a singular carbon source at 50 °C incubation temperature [61]; and the development of a prototype machine, which distills oil from PE [62]. It has also been found that under thermo- and photo-oxidation conditions, *Acinetobacter* sp. 351 can degrade lower molecular weight of PE oligomer yielding products such as ester, dicarboxylic acids, carboxylic acid, aldehydes, alcohols, keto-acids, lactones, ketones, alkanes and alkenes [63]. It has further been found and demonstrated that the *Plodia interpunctella* moth larvae could lunch on PE, and degrade it to the extent of causing reduction of tensile strength 50%, mass by 10% and molecular weights of the resulting polymers by 13% [64, 65].

Scientifically, recycling is “the collection and, often reprocessing of discarded materials for reuse.” Materials for recycling include those employed as consumer products and those used in manufacturing processes. Recycling causes reduction in public health risks, conserve natural resources through pollution reduction and extend the useful period of landfills. It also helps recover material matter and energy used to produce the material to some extent. Though not all plastics can be recycled, PE used in the production of sachet water bags is sensitive to the technique. Once sorted from other plastic materials, PE can be recycled to reproduce consumer and manufacturing products. Though recycled materials are lower in value and quality than the parent material alleviation of pollution, environmental cost, realization by the public of cost of environmental externalities and public health cost of plastics more than compensate for these limitations. As one author puts it “if we want to really clean up our mess, the most effective, cheapest strategy is to prevent the plastic from getting there in the first place!” In this regard, we can prevent larger quantities of plastic waste from remaining in the environment through recycling at any particular point in time.

From the foregoing it can be observed that one of the best techniques of disposing plastics in general and PE in particular is to adopt a hybrid approach. Studies have evidenced that composite plastic-natural fiber material can be developed with the ability to meet both recycling and biodegradation criteria. The result is a reduction of plastic materials that enter the environmental space at any given time; and availability of bio-organisms that can consume plastic materials that enter the environmental space as well. The natural fiber degrades naturally in the environment. Thus in terms of material mass and volume, the quantum of plastics exposed to the environment could be reduced by the volume and material mass that is added to the PE-natural fiber composite material. It appears that the major challenge delaying the progress of the application of plastic natural fiber composite material is the type of fiber or plant type and the production technique that can enhance the properties of PE. The hydrophobic characteristic of resulting composites is also another major challenge. Future studies when directed in dealing with these challenges shall go a long way to encourage the use and application of plastics such as PE with insignificant detriment to both current and future generations.

References

[1] North, E. J., and Halden, R.U. Plastics and environmental health. The road ahead. Rev Environ Health 2013, 28(1), 1–8. doi: 10-1515/reveh-2012-0030.

[2] American Chemical Society. 1993. Bakelite: The world's first synthetic plastic. Chemical Landmarks program of the American Chemical Society. Retrieved June 18, 2017, from https://www.acs.org/content/acs/en/education/whatischemistry/landmarks/bakelite.html.

[3] Ragaert, K., Delva, L., and Van Geem, K. Mechanical and chemical recycling of solid plastic waste. Waste Manage 2017, 69, 24–58.

[4] Market Study.2014. Polyethylene – LLDPE. 2nd Edition. November 2014.

[5] GLOBE NEWSWIRE. 2018. Zion Market Research, Global Polyethylene Market Will Reach USD 215 Billion by 2024: Zion Market Research. Retrieved 2 February, 2019, from https://www.linkedin.com/company/zion-market-research).

[6] Cheabu, B.S.N., and Ephraim, H.H. Sachet water quality in Obuasi, Ashanti Region, Ghana. J Biol, Agri Health Care 2014, 4(5), 37–42.

[7] Von Pechmann, H. Ueber diazomethane and nitrosoacylamine. Berichte der Deutschenchemischen. Gezellschaftzu Berlin 1898, 31(2640–2646), 2643.

[8] Bamberger, E., and Tschirner, F. 1990. On the effect of diazomethane on B-arylhydroxylamine (Ueber die Einwirkung van Diazomethan auf B-Arylhydorxylamme. Berichte der Deutschenchemischen Gesellschaftzu Berlin, 33: 955–959, Winnington History in the Making (2006). This is Cheshire. 23 August 2006. Archived from the original on 21 January 2010.

[9] Hoff, R., and Mathers, R.T. Review of Phillips Chromium Catalyst for Ethylene polymerization, Hoff, ray: mathers, Robert T, Handbook of Transition metal polymerization catalysts, 2010, John Wiley & Sons. doi: 10.00/9780470504437.ch10.

[10] Kaiser, W. KunststoffchemiefürIngenieare von der synthesebis zur Anwendung, 2011, 3ed, Hanser, München.

[11] Whiteley, K.S., Heggs, T.G., Kock, H., Mawer, R.L., and Immel, W. Polyolefins. Ullmann's Encyclopedia of Industrial Chemistry, 2005, Wiley-VCH, Weinheim. doi: 10.1002/14356007.a21487.

[12] Chung, C.I. Extrusion of Polymers: Theory and Practice, 2010, 2nd, Hanser, Munich.

[13] Market Study 2012. Polyethylene – HDPE, Cersana Research, May 2012.

[14] Trucost 2016. Plastics and Sustainability: a valuation of environmental benefits, costs and opportunities for continuous improvement, American Chemistry Council.

[15] Wiltz, J. 2017. Plastic bag pollution facts. June 13, 2017, from http://www.epa.gov/climatechange/waste/downloads/ICF_Memo_Carbon_Seqeustration_in_landfills-pdfhttp://www.livestrong.com/article/1599961-plasticbag.

[16] Ryan, D.B. 2017. Environmental problems with plastic bags. Retrieved Jun 13, 2017, from http://www.livestrong.com/article/216666-why-are-plastic-bags-so-bad-for-the-environemtn/

[17] New, M. 2017. Why are plastic bags so bad for the environment? Retrieved Jun 13, 2017, from http://www.livestrong.com/article/216666-why-are-plactic-bags-so-bad-for-the-envirnment/
[18] Disadvantages of plastic Bags (2017). Retrieved from http://lifestyle.iloveindia.Com/lounge/.
[19] Lee, K. 2014. The negative effects of using plastic drinking bottles. Retrieved Jan 29, 2014, from http://www.livestrong.com/articles/183101-the-negaive-effects-of-using-plastic-dringking-bottles/
[20] Law, K.L., Moret-Ferguson, S., Maximenko, N.A., Proskurowsky, G., Peaccock, E.E., Hanger, J., and Reddy, C.M. Plastic accumulation in the North Atlantic subtropical Gyre. Science 2010, (329), 1185.
[21] Container Recycling Institute. December 21, 2015. Container Recycling Institute says beverage container recycling and sales increase in California. Retrieved Dec 03, 2018, from https://www.recyclingtoday.com/article/cri-analyzes-california-crv-data/
[22] Frost, S. 2015. Disadvantages of using plastic bags, Retrieved June 29, 2018, from. http://www.livestrong.ocm/article/156070-disadvantages-of-using-plastic-bags.
[23] Edwards, C. 2017. Harmful effects of plastic bottles. Retrieved June, 2017, from http://www.livestrong.com/article/183101-the-negative-effects-of-uisingg-plastic-drinking-bottles/.
[24] Moore, J. Synthetic polymers in the marine environment: A rapidly increasing, long-term threat. Environ Res 2008, (108), 131.
[25] Jambeck, J.R., Geyer, R., Willllcox, C., Seigler, T.R., Perryman, M., Andrady, A. et al. Plastic waste inputs from land into the ocean. Science 2015, 347(6223), 768–771.
[26] Baffour-Awuah, E. Health implications of polluted tilapia consumption – The perception of Fosu Lagoon fishermen in Cape Coast, Ghana. J Environ Earth Sci 2014, 4(10), 78–86.
[27] Gregory, M. Environmental implications of plastic debris I marine settings –Entanglement, ingestion, smothering, hangers-on, hitch-hiking and alien invasions. Phil Trans R Soc B 2016, 364, 1526.
[28] Calafat, AM., Kuklenyk, Z., Reidy., J.A., Causdill., S.P., Ekong, J., and Needham, L.L. Urinary concentrations of bisphenol A and 4-nonylphenol in a human reference population. Environ Health Perspect 2005, 113(4), 391–395. (PubMeu:15811827).
[29] Romao, S.N., Edwards, H.E., Souder, J.P., Ryan, K.J., Cni, X., and Gorelick, D.A. protein-coupled estrogen receptor regulates embryonic hearth rate in zebrafish. PLOS Genet 2017, 13, 10. doi: 10.1371/journal.pen.1007069.
[30] Diamanti-Kandarakis, E., Bourguignon, J., Giudice, L.C., Houser, R., Prins, G.S., Soto, A.M. Et al. Endocrine-disrupting chemicals: An endocrine society scientific statement. Endocri Rev 2009, 30(4), 293–342. [PubMed: 19502515].
[31] Halden, R.U. Plastics and health risks. Annu Rev Public Health 2007, 31(1), 179–194. [PubMed: 20070188].
[32] Richter, C.A., Birnbaum, L.S., Farabollini, F., Newbold, R.R., Rubin, B.S., Tasness, C.E. Et al. In vivo effects of bisphenol A in laboratory rodent studies. Repro Toxicol 2007, 24(2), 199–224. [PubMed:10856020].
[33] Safe, S.H. Endocrine disruptors and human Health is there a problem? An update. Env Health Perspect 2000, 108(6), 487–493. [PubMed: 10856020].
[34] Onundi, Y., Drake, B.A., Malecky, M. A., de Nardo, M.R., Mills, S.K., and Pyabor, A.D. A multidisciplinary investigation og the technical and environmental performances of TAML/Peroxide elimination of bisphenol a compounds from water: destruction, oligomerisaton, mechanism, end product toxicity, and applications. Green Chem 2017. doi: 10.1039/c7gc01415e.
[35] University of Cincinnati. 2008. Plastic bottles release potentially harmful chemicals (Bisphenol A) after contact with hot liquids. Science Daily, 4 February 2008. Retrieved June 13, 2017, from www.sciencedaily.com/releases/2008/01/080130092108.htm

[36] Manshack, L.K., Conard, C.M., Bryan, S.J., Deem, D.K., Holliday, N.J., Bivens, S.A. et al. Transcriptomic alterations in the brain of painted turtles (chrysemyspicta) developmentally exposed to bisphenol A or ethinyl estradiol. Physiol Genomics 2017, 49(4). doi: 10.1152/physiolgenomics. 00103.2017.
[37] Kinch, C.D., Ibhazehiebo, K., Jeong, J.H., Habibi, H.R., and Kurrasch, D.M. Low-dose exposure to bisphenol A and replacement bisphenol S induces precocious hypothalamic neurogenesis in embryonic zebrafish. PNAS 2015. January 12, 2015. doi: 10.1073/pnas. 1417731112.
[38] Endocrine Society. 2014. BPA exposure during fetal development raises risk of precancerous prostate lesions later in life. ScienceDaily, Jun 23, 2014. 222.sciencedaily.com/releases/2014/06/140623091955.htm>.
[39] Federal Register. 2012. The Daily Journal of the United States Government. Indirect Food Additives: Polymers. Retrieved June 29, 2018 from http://federalregister.gov/articles/2012/07/17/2012-17366/indirect-food-additives-polymerSHP.
[40] US Food and Drug Administration. 2015. Bisphenol A (BPA: Use in food contact Application. http://www.epa.gov/climatechange/waste/downloads/ICF_Memo_Carbon_Seqeustration_in_landfills-pdfhttp://www.fda.gov/newsevents/publichealthfocus/ucm064437.htm
[41] Singh, L., and Li, S.S. Bisphenol A and phthalates exhibit similar toxicogenomics and health effects. Genetics 2012, 494(1), 85–91. [PubMed:22173104].
[42] Houser, R., Meeker, J.D., Duty, S., Silva, M.J., and Calafat, A.M. Altered semen quality in relation to urinary concentrations of Phthate monoester and oxidative metabolites. Epidemiology 2006, 17(6), 682. [PubMed:17003688].
[43] AZ Chemistry. 2017. 15 list of chemicals in plastic – Properties-dangers. 17 July 2017. Azchemistry.com.
[44] Habert, R., Muczynski, V., Grisin, T., Moison, D., Messiac, R., Frydman, A.B. et al. Concerns about the widespread use of rodent models for human risk assessments of endocrine disruptors. Reproduction 2014, 147, 4. doi: 10.1530/REP-13-0497.
[45] Ayamba, A.A. Assessment of phthalate migration in polyethylene food packages: A case study in Ghana, 2015, University of Ghana, Accra, Retrieved from. ugspace.ug.edu.gh.
[46] Xu, Q., Yin, X., Wang, M., Wang, H., Zhang, N., Shen, Y. et al. Analysis of Phthalate migration from plastic containers to packaged cooking oil and mineral water. J Agric Food Chem 2010, 58(21), 11311–11317.
[47] Lule, F., Kigozi, J.B., Ssempala, C., and Banadda, N. Development and evaluation of models for predicting chemical contaminant immigration in foods, 2012, Makerere University Press. uganda.repostitory.ru.forum.
[48] Beach, E. 2015. Health effects of plastic water bottles. Dec 28, 2015. http://www.livestrong.com/article/131685-health-effects-plastic-water-bottles/
[49] Downey, L. 2015. Are there effects of bottled water? Retrieved Dec 3, 2018 from. http://www.livestrong.com//article/154123-bottled-water-side-effects/.
[50] Martinez, E. 2015. What are the danger to boiling food in plastic bags? October 06, 2015. http:///www.livestrong-com/article/176349-what-are-the-dangers-of-boiling-food-in-plastic-bags/
[51] Choe, S.V., Kim, S.J., Kim, H.G., Choi, Y., Lee, H., and Kim, Y. Evaluation of estrogenicity of major heavy metals. Sci Total Environ 2003, 312(1–3), 15–21.
[52] Obiakor, M.O., Tighe, M., Pereg, L., and Wilson, S. Bioaccumulation, trophodynamics and ecotoxicity of antimony in environmental freshwater food webs. Crit Rev Environ Sci Technol 2018, 47(22), 1–51. doi: 10.1080/10643389.2017.1419790.
[53] Guo, W., Fu, Z., Wang, H., Song, F., Wu, F., and Giesy, J.P. Environmental geochemical ad spatial/temporal behavior of total and speciation of antimony in typical contaminated aquatic environment from Xikuangshan, China. Microchem J 2018, 137, 181–189.

[54] Moazzen, M., Mahvi, A.H., Shariatifar, N., Khaniki, G.J., Nazmara, S. S., Alimohammed, M. et al. Determination of phthalate cid esters (PAEs) I carbonated soft drinks with MSPE/GC-MS method. Toxins Rev 2017, 0(0), 1–8.
[55] US Environmental Protection Agency. 2015. Carbon Sequestration in Landfills- US environmental Production Agency. Retrieved June 13, 2017, from http://www.epa.gov/climate change/waste/downloads/ICF_Memo_Carbon_Seqeustration_in_landfills-pdf.
[56] Hopewell, J., Dorak, R., and Kosior, E. Plastics recycling challenges and opportunities. Phil Trans R Soc B 2009, 364(1526), 2115–2126. [OybNedL18527958].
[57] Lisk, D.J. Environmental implications of incineration of municipal solid-waste and ash disposal. Sci Total Environ 1988, 74, 39–66. [PubMed:3065938].
[58] ASTM Standards. 1999. ASTM Standard D 6400-99 Standard Specification for Compostable Plastics, 1999. https://www.astm.org. Standards & Publications.
[59] Chanprateep, S. Current trends in biodegradable polyhydroxylalkanoates. J BiosciBioeng 2010, 110(6), 621–632. [PubMed: 20719562].
[60] Canada World. Nov 12, 2008. WCI student isolates microbe that lunches on plastic bags. Retrieved 18 June, 2017, from https://hilaryfarlow.wordpress.com/…/wci-student-isolates-mi crobe-that-lunches-on-pl…
[61] Hadad, D., Geresh, S., and Sivan, A. Biodegradation of polyethylene by the thermophilic bacterium Brevibacillus Borstelensis. J Appl Microbiol 2005, 98(5), 1093–1100. doi: 10.111/ j.1365-2672-200.02553,x.
[62] Nguyen, T. 17 February 2011. New invention turns plastic bags into oil. Retrieved from smart planet.com.
[63] Tokiwa, Y., Calabia, B.P., Ygwu, C.U., and Aiba, S. Biodegradation of plastics. Int J Mol Sci 2009, 10(9), 3722–3742. doi: 10.3390/ijms10093722.
[64] Balster, L. January, 2015. Discovery of plastic-eating bacteria may speed waste reduction. Retrieved from www.fondriest.com.
[65] Yang, J., Yang., Y., Wu, W.M., Zhav, J., and Jiang, L. Evidence of polyethylene biodegradation by bacterial strains from the guts of plastic-eating wax worms. Environ Sci Technol 2014, 48 (23), 13776. doi: 10.1021/es504038a-.

Index

3-Point Bend Test 87

Adhesive Material 87
Anisotropy 34, 36
Anisotropy 26
ANOVA 90, 98, 100, 126, 138, 143, 148, 151
ANSYS 5, 7, 11, 12, 14, 101, 111, 113, 115, 117
Aspect Ratio 29, 34, 71, 76, 125, 127, 129, 130, 131

Bio-Adhesive 90, 91, 95, 98, 104, 106
Bio-Based Plywood 87
Bioplastic 57, 60, 181

Cavity Shape 36
Composite 25, 26, 37, 38, 39, 45–56, 57–80, 89, 94, 104, 106, 107, 109–118, 121–132, 143, 171, 173, 182
Composite Formation 63, 77, 79
Composite Performance 40, 63, 109, 143
Composite Properties 67, 79, 121, 130
CREO Parametric 7, 11
Cure Kinetics 46, 47, 50, 51, 52, 56, 111
Cure Viscosity 46, 56

Darcy's Law 110, 112, 128
Design 25–40, 45, 46, 77, 90, 92, 101, 110, 111, 116, 117, 124, 137, 140, 143, 160
Digital Software 39

Environmental Pollution 63, 79, 169, 175, 180
Epoxy 58, 59, 67, 76, 79

FESEM 90, 95, 98, 106
Fiber Concentration 29, 30, 129, 130
Fiber Orientation 25, 67, 123, 125, 130
Fiber Properties 63
Fiber Suspension 34
Fibreboard 87
Field Emission Scanning Electron Microscopy 95
Finite Element Analysis 3, 6, 8, 12, 20, 101, 104
Flexural Strength 71, 72, 73, 75, 76, 90, 104
Formaldehyde 58, 67, 90, 95, 98, 101, 103, 106
Formaldehyde Based Glue 90
FRP 3, 5

Gate Position 30, 125

High Stiffness 34
High Strength-To-Weight Ratio 34
Hybrid Composites 57, 69, 79

injection Molding 25–40, 121–132, 137–152
Injection Strategy 45, 111, 117, 118

L9 98
Laser Transmission Welding 154, 156, 163
Leaf Spring 17–32
Liquid Composite Moulding 109
Load Versus Displacement 87

Manufacturing 3, 18, 46, 59, 60, 76, 90, 91, 98, 99, 106, 110, 111, 121, 131, 138, 153, 170, 174, 179, 182
Mechanical Properties 10, 12, 38, 39, 63, 76, 124, 129, 130, 139, 172, 173
Modulus of Rupture 90, 92, 93, 96, 98, 100, 106
Mold 25–40, 110, 111, 112, 114, 115, 117, 118, 122, 123, 124, 126, 128, 138
Mold Design 30, 116, 137
Moldflow 34, 36, 37, 39, 131, 138, 139, 141, 144
Moldflow Insight 34, 36, 37, 39, 131
Mono Composite Carbon 3
Mould Filling Simulations 109, 112, 113, 115, 116

Nano Composites 3
Naturel Fibers 121

OPF 57–80
Optimization 45, 137–152, 160, 162, 163
Orientation 25–40, 68, 69, 76, 122, 124, 125, 130
Orthogonal Array 98, 106, 138, 141, 156, 163

Parameters 6, 9, 11, 14, 16, 17, 26, 28, 29, 38, 46, 52, 90, 93, 98, 100, 106, 111, 116, 125, 126, 131, 132, 137–152, 156, 157, 158, 161, 162, 164
Parametric Analyses 163
Particleboard 94

https://doi.org/10.1515/9783110655049-011

Performance Prediction 34
Plastic-Biofiber Composites 169
Plastics 58, 66, 79, 153, 154, 155, 170, 174, 175, 176, 177, 178, 179, 180, 181, 182
Plywood 89
Polyethylene 58, 67, 72, 77, 126, 169–183
Polymer 3, 25, 27, 28, 29, 31, 34, 47, 57–80, 109, 121, 122, 123, 143, 155, 182
Porous Flow 109
Pressure 20, 27, 33, 90, 92, 93, 98, 100, 110, 111, 112, 113, 114, 116, 117, 118, 124, 126, 128, 131, 138, 140, 144, 148, 151, 154, 156, 157, 158, 162, 163, 171, 175
Process Parameters 90, 94, 98, 100, 113, 138, 140, 141, 143, 149, 151, 157

REM 3D Software 39
Resin Exotherm 45
Resin Gelation 45, 46, 48
Resin Transfer Moulding 23, 45, 109

Self-Weight 34
Short-Fiber-Reinforced Thermoplastics 101, 104
Shrinkage 27, 50, 122, 124, 126, 127, 131, 138, 140, 143, 144, 148, 149, 151, 152
SMA 90, 91, 93, 95, 96, 98, 100, 101, 104, 106
Soy Meal Adhesive 90, 91, 94, 95, 97, 100, 101, 106
STAAD-Pro Software 101, 107
Stress-Stain Curves 87

Taguchi 90, 98, 99, 106, 137–152, 153–164
Taguchi Quality Loss Function Analysis 158, 163
Temperature 28, 33, 47, 48, 49, 50, 51, 52, 56, 58, 62, 69, 73, 90, 92, 95, 98, 100, 106, 111, 117, 122, 126, 131, 138, 140, 144, 148, 151, 154, 171, 177, 179
Thermoplastics 26, 39, 58, 73, 125, 130, 154
Toxic 175, 176, 179, 180

Volumetric Shrinkage 127, 131, 137–152

Warpage 123, 126, 131, 137–152
Wind Loads 101, 104

X-Ray Diffraction 95
XRD 90, 95, 97, 98, 106

Bei Fragen zur Produktsicherheit wenden Sie sich bitte an:
If you have any questions regarding product safety,
please contact:

Walter de Gruyter GmbH
Genthiner Straße 13
10785 Berlin
productsafety@degruyterbrill.com